PREPARING TEACHERS TO TEACH THE STEM DISCIPLINES IN AMERICA'S URBAN SCHOOLS

ADVANCES IN RESEARCH ON TEACHING

Series Editor: Volumes 1–11: ***Jere Brophy***
Volumes 12–29: ***Stefinee Pinnegar***

Recent Volumes:

Volume 19:	From Teacher Thinking to Teachers and Teaching: The Evolution of a Research Community
Volume 20:	Innovations in Science Teacher Education in the Asia Pacific
Volume 21:	Research on Preparing Preservice Teachers to Work Effectively with Emergent Bilinguals
Volume 22:	International Teacher Education: Promising Pedagogies (Part A)
Volume 22:	International Teacher Education: Promising Pedagogies (Part B)
Volume 23:	Narrative Conceptions of Knowledge: Toward Understanding Teacher Attrition
Volume 24:	Research on Preparing Inservice Teachers to Work Effectively with Emergent Bilinguals
Volume 25:	Exploring Pedagogies for Diverse Learners Online
Volume 26:	Knowing, Becoming, Doing as Teacher Educators: Identity, Intimate Scholarship, Inquiry
Volume 27:	Innovations in English Language Arts Teacher Education
Volume 28:	Crossroads of the Classroom: Narrative Intersections of Teacher Knowledge and Subject Matter
Volume 29:	Culturally Sustaining and Revitalizing Pedagogies
Volume 30:	Self-study of Language and Literacy Teacher Education Practices
Volume 31:	Decentering the Researcher in Intimate Scholarship: Critical Posthuman Methodological Perspectives in Education
Volume 32:	Essays on Teaching Education and the Inner Drama of Teaching: Where Biography and History Meet
Volume 33:	Landscapes, Edges, and Identity-Making
Volume 34:	Exploring Self Toward Expanding Teaching, Teacher Education and Practitioner Research

ADVANCES IN RESEARCH ON TEACHING, VOLUME 35

PREPARING TEACHERS TO TEACH THE STEM DISCIPLINES IN AMERICA'S URBAN SCHOOLS

BY

CHERYL J. CRAIG
Texas A&M University, USA

PAIGE K. EVANS
University of Houston, USA

DONNA W. STOKES
University of Houston, USA

emerald PUBLISHING

United Kingdom – North America – Japan
India – Malaysia – China

Emerald Publishing Limited
Howard House, Wagon Lane, Bingley BD16 1WA, UK

First edition 2021

British Library Cataloguing in Publication Data
A catalogue record for this book is available from the British Library

ISBN: 978-1-83909-457-6 (Print)
ISBN: 978-1-83909-456-9 (Online)
ISBN: 978-1-83909-458-3 (Epub)

ISSN: 1479-3687 (Series)

Printed and bound by CPI Group (UK) Ltd, Croydon, CR0 4YY

ISOQAR certified Management System, awarded to Emerald for adherence to Environmental standard ISO 14001:2004.

ISOQAR REGISTERED
Certificate Number 1985
ISO 14001

INVESTOR IN PEOPLE

CONTENTS

LIST OF FIGURES AND TABLES

ABOUT THE CONTRIBUTORS

Bobby Abrol, PhD, studied in the Department of Curriculum and Instruction at the University of Houston, Texas. After her 2016 graduation, she relocated to India to work in the development sector. Currently, she is working with Tata Trusts as a Lead in the "Developing Teachers" strand of the Education portfolio. Her work provides dual opportunities to engage with both the policy makers at the Ministry of Education level and to understand the ground realities through the Trusts' work in different States.

Cheryl J. Craig, PhD, is a Professor, the Houston Endowment Endowed Chair for Urban Education, and the Program Lead for Technology and Teacher Education at Texas A&M University. She is an American Educational Research Association (AERA) Fellow, a recipient of AERA's Division B (Curriculum) Outstanding Lifetime Achievement Award and a winner of AERA's Michael Huberman Award for Outstanding Contributions to Understanding the Lives of Teachers. The International Study Association on Teachers and Teaching (ISATT) has bestowed on her its highest honor: the ST^{2}AR Award for Significant and Exemplary Contributions through Research, Teaching and Professional Service in the International Field of Teaching and Teacher Education.

Paige K. Evans, Ed.D, is the Associate Director and Clinical Professor for the STEM teacher preparation program, *teach*HOUSTON, in the Department of Mathematics at the University of Houston (UH) where she teaches and oversees courses in preservice STEM teacher education. Her research interests include STEM education, STEM teacher preparation and culturally responsive pedagogy. She is an American Physical Society Physics Teacher Education Coalition Fellow and a recipient of the UH Teaching Excellence Award, the UH Group Teaching Excellence Award, and the NSM John. C. Butler Teaching Excellence Award. She recently served as the president of the UTeach STEM Educators Association.

Tenesha Gale, PhD, completed her studies in the Department of Teaching, Learning and Culture, Texas A&M University. Her research focuses on community changes from suburban to urban and how those changes affect education for Black, Brown, and low socioeconomic students. She is currently teaching science in the Humble Independent School District and continuing her research.

Xiao Han, Ed.D, is a postdoctoral researcher in the Department of Teaching, Learning and Culture at Texas A & M University. She majored in Instructional Technology and received her master's and doctoral degrees from George Mason

University and the University of Houston, respectively, and was the Director of Online Masters' Degree program at St. Thomas University, Houston. Her interests focus on narrative inquiry, instructional technology and design, teacher education, online learning, and educational reform.

Jing Li, PhD, completed her program of studies at the College of Education and Human Development, Texas A&M University. She currently works as a Research fellow in the Faculty of Education, East China Normal University. Her research focuses on student experiences; teachers' personal–professional development in urban and rural schools; teacher knowledge communities; teacher identity; teacher retention in high-needs areas; and narrative inquiry/digital narrative inquiry.

Mariam Manuel, PhD, is an Instructional Assistant Professor for the STEM teacher preparation program, *teach*HOUSTON in the Department of Mathematics at the University of Houston. She teaches undergraduate and graduate coursework in STEM education. Her research interests include STEM teacher education, engineering design education, and culturally responsive pedagogy.

Leah McAlister-Shields, Ed.D, is a lecturer and faculty advisor for the *teach*HOUSTON Program in the Department of Mathematics at the University of Houston. Dr McAlister-Shields has served on several NSF grant-funded projects that support STEM college student success. Her research and teaching interests include college student sociocultural capital, matriculation, and persistence of underrepresented STEM majors and culturally responsive teaching in post-secondary settings.

Jeffrey Morgan, PhD, is an Associate Provost and Professor of Mathematics at UH. He is responsible for encouraging online program development, identifying instructional technologies, and overseeing software design, LMS administration, and central instructional design. Morgan is also the founder and director of the Center for Academic Support and Assessment and the co-founder and co-director of *teach*HOUSTON, HoustonACT, and the Math Contest. Previously, Morgan was the Department Chair of Mathematics and is an active teacher and researcher.

Ha Nguyen is a third-year PhD student in the Higher Education Leadership and Policy Studies at the University of Houston (UH). Nguyen's research interests focus on college STEM student's success and STEM teacher education. Nguyen is currently a research and teaching assistant for UH's *teach*HOUSTON program. Prior to enrolling in her PhD program, Nguyen spent over 15 years working in the development sector in Vietnam. She served as a Monitoring and Evaluation Manager for the US Agency for International Development (USAID) in Vietnam.

Michele Norton, PhD, completed her studies in the Department of Teaching, Learning and Culture, Texas A&M University, College Station, TX. Currently, she is a Post-Doctoral Associate in the Department of Electrical Engineering at Texas at A&M University. She wrote a Texas A&M T3 Grant that funded the study of Emotional Intelligence (EI) of engineering students. Norton's research interests are in STEM education, engineering design process, design thinking, creativity, coaching, and social and emotional learning. She is currently working on two NSF-funded grants aimed at developing teachers and leaders in STEM Education.

Donna W. Stokes, PhD, is a Professor in the Physics Department and Associate Dean for Undergraduate Affairs and Student Success in the College of Natural Sciences and Mathematics at University of Houston (UH). Her scientific research focuses on understanding the structural/optical properties of semiconductors. Her education research focuses on physics/STEM student success and teacher education. She received her PhD at UH and was a postdoctoral researcher at the Naval Research Laboratory. She is an American Physical Society Physics Teacher Education Coalition Fellow and has been a recipient of a NSF Early Career Award, an UH Excellence in Group Teaching Award, and the Provost's Faculty Advising Award.

Rakesh Verma, PhD, is a Professor in the Department of Computer Science in the College of Natural Sciences and Mathematics at the University of Houston. He has received the University of Houston's Lifetime Faculty Award for Mentoring Undergraduate Research. This career award, given to one faculty member per year, acknowledges a professor's exceptional efforts over the years in demonstrating a commitment to undergraduate education and research.

Gang Zhu, PhD, is an Associate Professor at the Institute of International and Comparative Education, East China Normal University. His expertise includes teacher education, urban education, and comparative education. He has published in such journals as *Urban Education*, *Cambridge Journal of Education*, and *Journal of Language, Identity, and Education*.

LIST OF CONTRIBUTORS

Bobby Abrol	Developing Teachers Program Lead, Tata Trust, India
Cheryl J. Craig	Professor, the Houston Endowment Endowed Chair for Urban Education, and the Program Lead for Technology and Teacher Education, Texas A&M University, Texas
Paige K. Evans	Associate Director and Clinical Professor, *teach*HOUSTON program, University of Houston (UH), Texas
Tenesha Gale	Science teacher in the Humble School District, Houston, TX
Xiao Han	Postdoctoral Researcher, Department of Teaching, Learning and Culture, Texas A&M University
Jing Li	Research Fellow, East China Normal University, Shanghai
Mariam Manuel	Instructional Assistant Professor, STEM teacher preparation program, *teach*HOUSTON, University of Houston, Texas
Leah McAlister-Shields	Lecturer and faculty advisor, *teach*HOUSTON Program, University of Houston, Texas
Jeffrey Morgan	Associate Provost and Professor of Mathematics, University of Houston, Texas
Ha Nguyen	Graduate Assistant in the College of Education at the University of Houston. Research Assistant with *teach*HOUSTON
Michele Norton	Post-Doctoral Associate in the Department of Electrical Engineering at Texas A&M University
Donna W. Stokes	Professor and Associate Dean, University of Houston (UH), Texas
Rakesh Verma	Professor, University of Houston, Texas
Gang Zhu	Associate Professor, East China Normal University, Shanghai

FOREWORD

*teach*HOUSTON: Interdisciplinary, Experiential STEM Reform

F. Michael Connelly

What is one to make of a book with the title *Preparing Secondary STEM Teachers to Teach in America's Urban Schools* with what appears in the Abstract to have the subtitle *teach*HOUSTON? One part is rather grand and universal and the other more specific and local. I tend to think of subtitles, stated or implied, as reliable content descriptors. But my expectations faded as page after page and chapter after chapter unfolded. To be sure, *teach*HOUSTON names a concrete geographically limited program. The book assesses the status of teaching and learning in the STEM fields and describes a specific program to address the worrying picture that emerges. The program has direct consequences and possibilities for the Houston area. But the book as a whole and its account of this program addresses critical educational issues worldwide and demonstrates a kind of interdisciplinary action research rarely seen. Both the program and its practical demonstration of interdisciplinary action provide models useful in other jurisdictions and for other interdisciplinary sets. Unlike some, perhaps most, reform-oriented books, this one does not move inexorably step by step from problem to program to solution. Instead, different entry points and perspectives come in and out of focus in different chapters. Late in the book, in Final Words section, it is observed that the volume might be thought of as a mélange. Readers accustomed to step-by-step reform accounts might benefit by jumping from the opening overview chapters to Final Words section to better prepare them for the rich, original work found throughout. This book has insights and learnings for, and beyond, the STEM fields. Readers with different interests will find the book filled with suggestion and insight. Some of these possible readings follow.

The book describes a curriculum reform program in more or less standard educational change terms. The text begins with a thoughtful account of the educational learning problem, a description of the reform initiative to address the problem, and short-term demonstrations of evidence suggesting the problem is on its way to being solved. There is much to be learned about the status of STEM education and what might be done about it in the book's pages. As a Canadian reader, I would have welcomed a slightly expanded global picture but the language used to address the problem is readily applicable to a larger landscape. But what grips my attention and makes this book special in the literature of STEM education and educational reform is that suffusing a more or less traditional educational reform structure are two initiatives that warrant two follow-up monographs: interdisciplinary collaboration and the link between reform experience and content. It is well known that educational reforms rarely outlast the input of reform support and, when they do, they fade and become invisible over time. I am struck by the thought that this book describes two qualities that have a chance to defy these reform "facts."

Joe Schwab (1960) who is featured throughout this book showed how scientists within disciplines adhere to particular forms of thought. Both he and the philosopher Kuhn (1970) pointed out that changing these formal ways of thinking constituted revolution within fields of inquiry. School education and the education of teachers are cross-disciplinary. The purpose is to educate persons, not advance inquiry in a discipline. It is reasonably well known that any curricular reform must be done in the context of all the other school subjects and disciplines. Add some science to the school day, subtract some geography. Thus, the STEM disciplines are inherently in conflict with one another in traditional educational reform. The reform brilliance of STEM is that it puts the disciplines on the same side. This, however, is where Schwab and Kuhn's insight into forms of disciplinary thinking comes into play. If it is revolutionary for a scientist within a discipline to confront new ways of thinking about his or her discipline, consider the problem of STEM with different disciplines at work, each with their own traditions of thought. Moreover, educational scholars rarely fit easily among those in the traditional disciplines, thereby immensely confounding the interdisciplinary mix. Add to the interdisciplinary mix the voice and action of school teachers and teacher educators whose ends in view are the education of persons in contrast to the advancement of a discipline. The resulting logical interdisciplinary stew is immensely complex. This frames the work of the complex *teach*HOUSTON STEM Project. There is a telling study by Seymour Fox (1972), one of Schwab's students, who sets up a curriculum deliberation study involving school-based educators and subject area academics. He found that the educators deferred to the academics such that this dimension of interdisciplinary planning was missing. But throughout this book remarkable things are described. Two academic university departments joined hands. Academic educators and school people joined in. Chapter by chapter, the reader is led through what I consider to be remarkable settings...mathematicians talking to scientists, professors learning from school teachers, teacher educators and curriculum specialists providing ideas, literature, and experiential research methods shaping the mix.

The picture that emerges is rich and borders on the edge of believability. I would love to see a follow-up manuscript that made the remarkable interdisciplinary qualities described herein the subject of inquiry. What does it take among the disciplines, and among educators and practitioners, to carry off a successful interdisciplinary collaboration? What difficulties were encountered? How were they conceptualized and solved? Is there any evidence that Schwab and Kuhn might have overdramatized forms of thought or, perhaps, that there are cultural shifts toward more organic ways of thinking? Were there, one wonders, members of the *teach*HOUSTON team who were influenced by Chinese or Asian thought, which might have influenced the willingness to work across disciplinary lines of thought?

The second standout quality of this book is what I earlier called the link between reform experience and content. The massive international curriculum reforms of the last century, broadly falling under the heading of post Sputnik mobilization, led to an educational industry of reform and reform study. One of the mostly unchallenged insights from that industry is that telling school people what to do does not work. Yet it is clear that people with ideas about the practical value of their disciplines or their work look for ways to implement their ideas. People with good ideas about teaching and learning are perennially trapped in the logistic web of the wise and knowledgeable teaching the unwise and unknowledgeable. Returning to Schwab and Kuhn, and speaking rather broadly, the form of thought is one that the philosopher McKeon (1952) called logistic. Work out good ideas, and figure out how to train people to use them. The general failure of this form of thought about school change led to elaborate, often quite sensitive and responsive, ways of implementing ideas. Turned upside down, School-Based Reform took hold. *teach*HOUSTON stands in between and has it both ways. On the one hand, the best disciplinary ideas are at play. On the other hand, advanced notions of experience and their role in learning and in research method are at work. I know of no other major educational reform program that utilizes experiential method and theory in the context of an ongoing disciplines-based reform. The idea that the disciplines can bring insight and make a difference to schooling while, at the same time, learning from participating school practitioners is a rare quality in the study of educational reform. The fact that these processes are sufficiently recognized by *teach*HOUSTON participants at all levels to warrant specific research methodology built on the uncontrolled quality of experience is special. As with everything described in these pages, this quality requires intellectual strength and leadership along with collaborative interdisciplinary spirit and action. Again, I urge the authors/participants to consider a follow-up manuscript in their intended three-book series to unpack this "secret sauce" quality so vital to understanding *teach*HOUSTON and the interdisciplinary practices and research embedded in it.

My final thought is that in reflecting on my remarks about this book I have not been altogether transparent about my knowledge of the intellectual and practical dynamics at play. I was a student of Schwab and spent much time thinking about many of the issues discussed in this book. Moreover, I have followed Professor Craig's work for many years and am aware of her international reputation in experiential school-oriented thought. I do not know with certainty that her Asian studies influenced the holistic interdisciplinary thinking evident in this book (Craig et al., 2018). But the shape of the *teach*HOUSTON project and of the book built around it is consistent with the philosophical and practical trajectory of Craig's work. There must be something special about the Houston educational environment that brought so many diverse participants together in this atypical interdisciplinary, experiential, collaborative journey.

REFERENCES

Craig, C., You, J. E., Zou, Y., Verma, R., Stokes, D., Evans, P., & Curtis, G. (2018). The embodied nature of narrative knowledge: A cross-study analysis of embodied knowledge in teaching, learning, and life. *Teaching and Teacher Education*, *71*, 329–340.

Fox, S. (1972). A practical image of the practical. *Curriculum Theory Network*, *10*, 45–57.

Kuhn, T. (1970). *The structure of scientific revolutions*. Chicago, IL: University of Chicago Press.

McKeon, R. (1952). Philosophy and action. *Ethics*, *62*(2), 79–100.

Schwab, J. (1960). What do scientists do? *Behavioral Science*, *5*, 1–27.

OVERVIEW OF THE BOOK

Cheryl J. Craig, Paige K. Evans and Donna W. Stokes

ABSTRACT

This chapter outlines the contents of Preparing Secondary STEM Teachers to Teach in America's Urban Schools. *The volume begins with an overview of the teach*HOUSTON *STEM teacher education program in Chapter 2 and is followed by an account of the collaboration that took place between a Physics professor and a teach*HOUSTON *Physics teacher educator and its impact on STEM teacher preparation in Chapters 3-4. Chapters 5 and 6 include discussions about formal and informal learning opportunities and include a narrative of a student's experience on how the Noyce Internship Institute contributed to their STEM teacher learning. In Chapters 7–9, readers learn about the influence of parents, teachers, and professors on students' entry into and decision to work in the STEM and/or STEM education field, with an emphasis on those in STEM teacher education. Chapter 10 highlights the value of scholarship grants; Chapter 11 addresses the growth and development of teach*HOUSTON, *the impact of the scholarships awarded to many of its students and traces where its graduates currently are teaching in order to demonstrate that teach*HOUSTON *has fulfilled its mission. The final chapter of the book reflects on teach*HOUSTON *as a secondary urban teacher education program and summarizes significant points that have led to its success.*

Keywords: STEM teacher preparation; secondary teacher education; urban teacher education; physics teacher preparation; National Science Foundation; *teach*HOUSTON

This book shines a spotlight on *teach*HOUSTON, a secondary education program that grew from a mere seed of an idea to a fully fledged science, technology, engineering, and mathematics (STEM) teacher education program with a faculty and staff all on its own. This first chapter, Chapter 1, which is authored by Cheryl

Preparing Teachers to Teach the STEM Disciplines in America's Urban Schools
Advances in Research on Teaching, Volume 35, 1–5

ISSN: 1479-3687/doi:10.1108/S1479-368720210000035001

J. Craig, Paige K. Evans, and Donna W. Stokes, sketches the different coauthored contributions to this volume and elucidates how each chapter builds on the others to offer a prismatic view of *teach*HOUSTON. Throughout this volume, readers will learn of *teach*HOUSTON's challenges and successes as a leading STEM teacher education program, in producing secondary teachers for America's urban schools.

Chapter 2, *Overview of the teach*HOUSTON *Program*, highlights the need for highly qualified STEM teachers in Texas, most especially in Houston, the fourth largest city in the United States. The strength of this chapter, authored by Paige K. Evans, Cheryl J. Craig, Donna W. Stokes, and Jeffrey Morgan, is that all the requisite pieces – the backbone of the program, so to speak – are painstakingly accounted for, which makes *teach*HOUSTON's key features visible and accessible to those who want to know more about it. Readers are also introduced to inquiry-based learning, student-centered instruction, and culturally responsive pedagogy. Additionally, the improvements made to the program over time are recounted. Most importantly, the collaboration between representatives of the Physics Department and the *teach*HOUSTON faculty is made known. This is especially informative because *teach*HOUSTON is housed in the Mathematics Department and has its offices located in the College of Education, which means the program's roots are truly cross-disciplinary.

Chapter 2 naturally leads to Chapter 3, *Collaboration Between a Physics Professor and a Physics Teacher/Teacher Educator*, which is coauthored by Donna W. Stokes, Paige K. Evans, and Cheryl J. Craig. Reduced to the essence, this chapter chronicles how a "deep collaboration" took shape between a Physics professor and a Physics teacher educator. While issues having to do with the cross-listing of courses for Physics and Physics Education students most certainly arose, the collaborators spoke in one voice to ensure that the rigor of the Physics and Physics teacher education coursework was never challenged again. This reinforced how both professors needed to work hand-in-hand at the forefront, otherwise *teach*HOUSTON and *teach*HOUSTON's scholarship students would risk being perceived as inferior – as opposed to being different – from pure Physics majors.

Authored by Paige K. Evans, Donna W. Stokes, and Cheryl J. Craig, Chapter 4, *A Narrative Inquiry Into Teaching Physics as Inquiry: One Teacher's Journey*, shares the lived experiences of undergraduate student whose pseudonym is Jason. The authors follow Jason through *teach*HOUSTON program and his subsequent school placements. The chapter fittingly ends with Jason becoming a cooperating teacher and mentor for other *teach*HOUSTON preservice teacher candidates. Here, the reader will see how the leadership of *teach*HOUSTON sustains the program through robust alumni participation, which also will be touched on in some of the chapters that follow. Also included in this chapter is the Physics as Inquiry course that Paige Evans developed and enacted. A critical underpinning of *teach*HOUSTON's curriculum, the new Physics course engages secondary preservice teachers in interactive, inquiry-based teaching pedagogies for Physics. The chapter not only shows how the course was designed to meet the needs of students of color but also how it fulfills content requirements in terms of credit hours for different STEM certification purposes. Physics majors and minors can obtain 7-12 Math/Physics certification, which allows teachers to teach both math and science; 7–12 Physical Science certification, which allows them to teach both Physics and Chemistry; and 7-12 Composite Science certification, which allows teachers to teach any Science.

In Chapter 5, readers learn about related activities of *teach*HOUSTON. Written by Donna W. Stokes and Paige K. Evans, the chapter, *Enhancing Preservice Teacher Preparation Through Formal and Informal Learning Experiences*, tells how *teach*HOUSTON's preservice teachers are afforded experiences to enhance their content development and teaching skills such as workshops, discipline specific courses, conferences, professional development opportunities, and paid internships. For example, interns become inducted as counselors and teaching assistants for a summer STEM camp for underserved middle school students for training in both formal and informal settings. Readers will also learn how topics varying from Bullying to the Growth Mindset are given front-stage attention through interactive experiences. Engagement in formal and informal activities, taught by both Physics and *teach*HOUSTON faculty and staff, allows students to be exposed to science as inquiry and project-based learning.

Chapter 6, *Examining the Impact of Informal Experiences on Preservice Teachers' Self-efficacy*, is one of several multiauthored chapters that follow. The chapter's first author is Paige K. Evans and her coauthors are Leah McAlister-Shields, Mariam Manuel, Donna W. Stokes, Cheryl J. Craig, and Ha Nguyen. The work highlights the impact of informal learning experiences for students pursuing STEM teaching careers at a time when considerable shortages of appropriately certified teachers exist in America's urban schools.

Chapters 7, 8, and 9 are a trilogy. Each deals with a particular influence on STEM students' decision to enter into and stay in STEM fields, including STEM teacher education. They illustrate how students' pursuing the STEM disciplines are shaped not only by their own experiences but also by those around them. Chapter 7 focuses on the influence of parents, Chapter 8 tells about the influence of teachers, and Chapter 9 is all about the influence of professors. These chapters also bring to the forefront the great diversity Houston has to offer. Although the book thus far has focused mainly on *teach*HOUSTON and teacher education, these chapters include who/why students pursue STEM teaching as well as other STEM fields such as Computer Science. This is important because computer science and technology are the next STEM content areas to be interwoven into the *teach*HOUSTON program through a recently awarded National Science Foundation (NSF) grant in 2020 to Principal Investigator, Paige Evans.

Chapter 7, *The Influence of Parents on Undergraduate and Graduate Students' Entering the STEM Disciplines and STEM Careers*, is used with the permission of the *European Journal of Education*. Authored by Cheryl J. Craig, Rakesh Verma (cybertechnology, cybersecurity), Donna W. Stokes, Paige K. Evans, and Bobby Abrol, it tells of the circumstantial and planned curriculum making that *teach*HOUSTON and computer science students experienced with their parents during their preschool and public school years. In this chapter, the home STEM experiences of three students, Katrina Roderick, Ryan Archer, and Sam Bernard (pseudonyms), are spotlighted.

Chapter 8, *A Tribute to "Unsung Teachers": Teachers' Influences on Students Enrolling in STEM Programs With the Intent of Entering STEM Careers* whose authors are Cheryl J. Craig, Paige K. Evans, Rakesh Verma, Donna W. Stokes, and Jing Li, published and used with the permission of the *European Journal of Teacher Education*, centers around teachers who left lasting imprints on the *teach*HOUSTON and computer science scholarship students. This chapter highlights three higher education students from very diverse backgrounds, pseudonyms being Joyce Harding (white female high poverty), Omid Kassem (immigrant to the United States), and Leon Mitchell (member of a historically underserved Black community second in size to Harlem).

The third paper of the trilogy is Chapter 9 of this volume, *The Influence of Professors on Students Enrolled in the STEM Programs with the Intent of Embarking on STEM Careers*. Authored by Jing Li, Paige K. Evans, Cheryl J. Craig, Donna W. Stokes, Rakesh Verma, and Gang Zhu, this chapter spotlights what the STEM students organically shared about learning from particular professors. The positive stories they told are ones that naturally spilled out because the STEM students needed to tell them, not because they were coaxed by the authors to reveal these storied nuggets. This makes it entirely fitting that this chapter is first-authored by Jing Li who herself was a student (albeit a doctoral one) at the time, as was Gang Zhu, another author of this multiauthored work.

Next in the lineup of chapters is *The Value of STEM Scholarship Grants to Undergraduate and Graduate Students Intending to Study the STEM Disciplines and Pursue STEM Careers*, which is Chapter 10. Jing Li is the first author with her coauthors being Cheryl J. Craig, Tenesha Gale, Michele Norton, Gang Zhu, Paige K. Evans, Donna W. Stokes, and Rakesh Verma. Seven stories of the impact that the NSF-funded scholarships had on the *teach*HOUSTON and computer science students are presented. The pseudonyms for the students are Omid Kassen, Anton Ivanov, Joyce Harding, Duong Pham, Sanjay Ritzvi, Kadeem Bello, and Tonya Goodson. This array of false names we chose also signals Houston's wide diversity as well as the diversity evident in *teach*HOUSTON and in the computer science programs at the University of Houston. Each of the seven students identified a different way that the STEM scholarship grants helped them. The experiences shared in this chapter are truly eye-opening and profound. Every STEM educator should read this chapter. Also, every philanthropist should think and make funding decisions, keeping individual student narratives such as the ones featured in this work in mind.

Our second to last chapter, Chapter 11, has to do with *Where Are the teach*HOUSTON *Preservice Teacher Candidates Now? Are They Still in the Urban Teacher Force*?, which is coauthored by Paige K. Evans, Mariam Manuel, Ha Nguyen, Donna W. Stokes, Cheryl J. Craig, Xiao Han, and Jeffrey Morgan (*teach*HOUSTON's codirector). This chapter explores a different kind of impact question: the quantitative query concerning how many students were impacted and where *teach*HOUSTON alumni are currently working. The chapter asks: Did *teach*HOUSTON produce more physics teachers than the Houston region previously had in the 10 years before the program began? Did those benefiting from *teach*HOUSTON mostly stay in the region and/or urban schools? Do those teachers continue to be reflective of Greater Houston's shifting diversity?

From Chapter 11, we shift seamlessly to the Final Word, which is Chapter 12. In Chapter 12, authors Cheryl J. Craig, Paige K. Evans, and Donna W. Stokes reflect backward on *teach*HOUSTON and address key points that led to its creation and its sustenance. The authors additionally discuss their personal development and share how their job responsibilities have changed in response to *teach*HOUSTON's growth and change.

OVERVIEW OF THE *teach*HOUSTON PROGRAM*

Paige K. Evans, Cheryl J. Craig, Donna W. Stokes and Jeffrey Morgan

ABSTRACT

*teach*HOUSTON *is a university-based secondary STEM teacher preparation program that addresses the critical need for highly qualified STEM teachers in Texas and across the country. STEM teachers are prepared through early and ongoing field-based teaching experiences and rigorous research-based instruction that integrates content and pedagogy provided by faculty members who have extensive teaching experience in public schools.* *teach*HOUSTON *serves the fourth largest city in the United States, along with its satellite communities and has many noteworthy features which are mapped in this chapter. Particular attention is paid to inquiry-based learning, student-centered instruction, and culturally responsive pedagogy as well as the improvements in the program based on the collaboration between physics and* *teach*HOUSTON *faculty.*

Keywords: Teacher education model; urban teacher education; science as inquiry; secondary STEM teacher preparation; culturally responsive pedagogy; *teach*HOUSTON

Providing a solid foundation in STEM is important in the education of future physicists, engineers, computer scientists, chemists, and medical doctors who can

*An earlier version of this updated chapter is: Craig, C., Evans, P., Stokes, D. & Bott, S. (2017). Attracting, preparing and retaining teachers in high need areas: A science as inquiry model of teacher education. In M. Peters, B. Cowie & I. Mentor (Eds.), *A companion to research in teacher education*. New York, NY: Springer Publishing.

Preparing Teachers to Teach the STEM Disciplines in America's Urban Schools
Advances in Research on Teaching, Volume 35, 7–26

ISSN: 1479-3687/doi:10.1108/S1479-368720210000035002

contribute to scientific discoveries. Indeed, as the economy of the United States shifts, having a strong STEM trained workforce is essential for the country to remain globally competitive. Over the next decade, our nation will need approximately one million new STEM professionals which can be met by increasing the number of students trained in STEM fields (Xue & Larson, 2015). Nevertheless, this will require STEM teachers who have the necessary content area knowledge and pedagogy to foster success and excitement in STEM at a time when there is a severe shortage of STEM teachers. According to the US Department of Education (2016), a persistent shortage of STEM teachers exists in the United States and the overall US production of STEM teachers fell 20% between 2009 and 2014. It is problematic that the United States is losing jobs to the global workforce because local students have been inadequately prepared for careers that require strong mathematics and science skills. Additionally, our public school system has not shown progress, especially in mathematics and science education (Chen, 2009).

Addressing content area knowledge and problems of pedagogy is particularly daunting in Texas, which has approximately five million students. It is difficult to believe that greater than 30% of middle school mathematics and science teachers are teaching out of their field and 13.3% and 28.7% of high school teachers are teaching out of their fields in mathematics and science, respectively (Texas Education Agency, 2017). This means these teachers instructed STEM classes without receiving adequate training and preparation. This is even more evident in the secondary schools in underserved areas of Houston which tend to employ the least qualified teachers and have the greatest shortages. This contributes to the achievement gap that Darling-Hammond (2011) termed an "opportunity gap." Nelson, Palonsky, and McCarthy (2009) suggest that underserved students are twice as likely to have uncertified teachers in comparison to their white peers. Figs. 1 and 2 illustrate the percentages of high school full-time equivalent school science and mathematics teachers in Texas who are assigned positions out of field in relation to the percentages of youths living in poverty.

Attracting highly qualified teachers to teach in high-poverty schools is problematic due to lack of funding, support, and resources. Also, good teachers according to American law are prepared in the content areas they are teaching. It has been shown that low-income students have only a 10% chance of having a good teacher in their K-12 education (Chenowith, 2010). Students from middle class homes have a support network which enables them to persist if taught by one or two weak teachers whereas high poverty students are less likely to recover. According to Villegas and Clewell (1998), these risks may be lessened by increasing the number of minority teachers as they may provide desperately needed role models for minority youth as they develop their own racial identities within an educational context. Students from low socioeconomic backgrounds may also benefit from the presence of minority teachers as they may foster successful academic outcomes in part to the contributions of minority teachers serving as role models for students of color (Dee, 2005; Pitts, 2007).

The urgent problem of preparing a sufficient number of qualified STEM teachers is matched only by the equally important need to retain STEM teachers

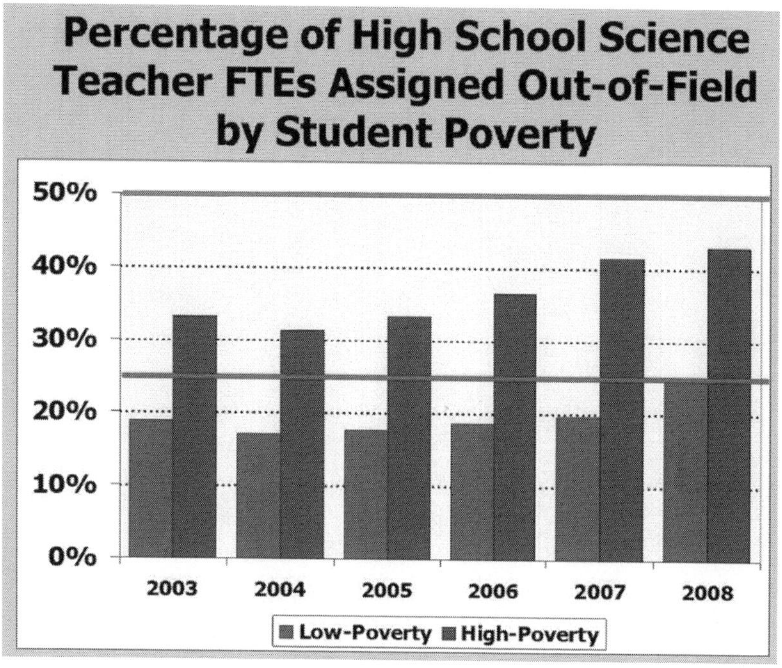

Fig. 1. Percentage of High School Science Teachers Assigned Out of Field in Texas from 1999 to 2008 by Student Poverty. *Source:* (Fuller, 2009).

in their careers, which is a challenge faced by most of the world's developed countries. Perhaps nowhere in the world is the teacher attrition problem more advanced than in the United States. For example, as early as 2009, 50% of Greater Houston's novice teachers (5 years or less of service) quit teaching by their fourth year of service. The city additionally has suffered from a serious "experienced teacher dropout rate" (National Public Radio, 2014). Craig (2014) contends that in 2014, in Houston's largest school district, the eighth largest in the nation, 80% of its teachers had five years or less of experience and 50% of its principals had five years or less service.

This increase in demand for STEM workers and shortage of STEM teachers warrants the importance of increasing numbers of qualified STEM teachers and providing them with critical professional support to help with retaining these teachers. Thus, a different model of STEM teacher preparation began at the University of Houston (UH) in 2007. The model, *teach*HOUSTON, is a replication of the UTeach program launched at the University of Texas at Austin. In this chapter, we feature the *teach*HOUSTON model for two important reasons. The first reason is that it addresses Greater Houston's teacher attrition rate, particularly where secondary content area teachers (mathematics, science, and special education) are concerned, which is among the most concerning issues in the United States and in the developed regions of the world. Second, the

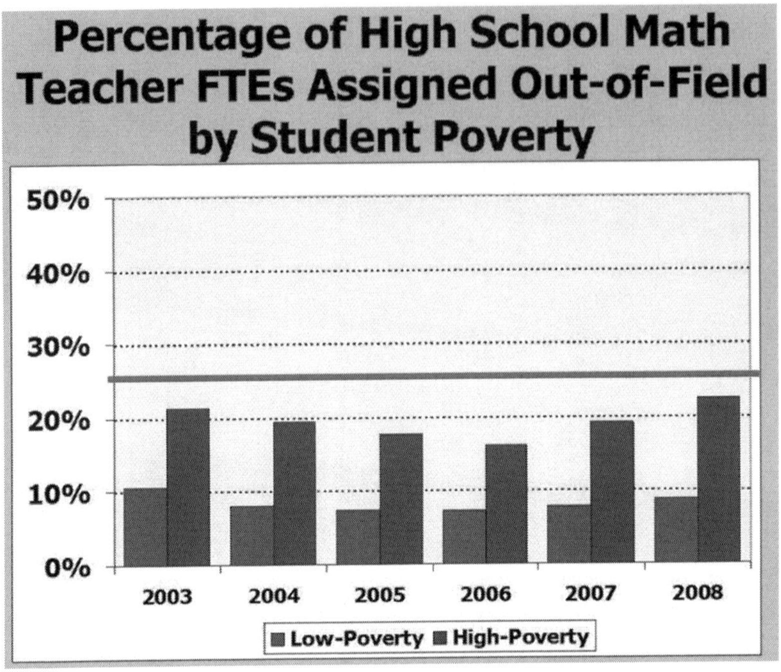

Fig. 2. Percentage of High School Mathematics Teachers Assigned Out of School in Texas from 1999 to 2008 by Student Poverty. *Source:* (Fuller, 2009).

present-day population mix in Texas will resemble that of the United States in 2040; therefore, it is fitting to create teacher education models that will contribute to the diversification of the teacher workforce and mirror that future population blend. Hence, it is important to show how *teach*HOUSTON is successfully attracting, preparing, and retaining secondary science teachers in what appears to be an against-the-odds situation.

This chapter begins by tracing the theoretical roots of teaching as inquiry and then discussing the contemporary view of science as inquiry. It continues by elucidating in fine-grained detail the *teach*HOUSTON program. Lastly, statistics over the increasing number of students served in Greater Houston by *teach*HOUSTON STEM teachers is shared followed by a figure that depicts *teach*HOUSTON's evolving approach to secondary STEM teacher preparation.

ORIGINS OF THE *teach*HOUSTON PROGRAM

Intellectual Lineage of Teaching as Inquiry and Science as Inquiry

In North America, the origins of teaching as inquiry trace to John Dewey, and the origins of science as inquiry trace to Dewey and Joseph Schwab. Dewey was the first to give mindedness to teachers and to see them as thinking, acting human beings. For Dewey (1938), teachers were a great deal more than vessels through whom others' codified knowledge prescriptions passed through. Dewey maintained that learning is deep rooted in the experiences a teacher/student enters into and the knowledge which arises from their inquiries. In Dewey's view, the inquiring mind is triggered by a perplexing experience or a discrepant event that leads the student/teacher to think reflectively and to engage in some type of action to resolve the problem. In this way, reflective thinking involves both the past and the future that students/teachers build upon from previous experiences and present knowledge to construct new knowledge that informs future experiences. Because of these baseline understandings, Dewey advocated for an experimental approach to science teaching. For him, the scientific process of inquiry was foundational to the discovery, use, and application of scientific knowledge.

Joseph Schwab, who was Deweyan in his orientation, advocated for students learning scientific concepts through inquiry and reinforced the teaching of science as inquiry in the field of education. Schwab's personal experiences as a scientist and curriculum theorist convinced him that the inquiry approach was the most defensible one. For Schwab, "scientific research has its origin, not in objective facts alone, but in a conception, a construction of the mind" (Schwab, 1962, p. 12). In his iconic treatise on *Teaching science as inquiry*, he went on to say that

> ...the treatment of science as [i]nquiry is not achieved by talk about science or scientific method apart from the content of science. On the contrary, treatment of science as [i]nquiry consists of a treatment of scientific knowledge in terms of its origins in the united activities of the human mind and hand which produce it; it is a means for clarifying and illuminating scientific knowledge. (p. 102)

It was not surprising, then, that Schwab was highly critical of textbooks that presented scientific facts as "rhetoric of conclusions" (p. 24) - that is, irrevocable truths. Such assertions, in Schwab's view, are only tentative stories because they remain truthful only until new discoveries are made.

Having sketched the roots of teaching and learning as inquiry and science as inquiry in place, we now turn to contemporary literature and policy documents supporting science taught as inquiry.

Science as Inquiry: A Current View

One of Dewey's most salient contributions to science education, as mentioned earlier, is the fact that he advocated for an experimental approach to science teaching, which was then reinforced by Schwab who similarly favored the approach. Together, their influence is imprinted in the national teaching standards and even in the Texas Essential Knowledge and Skills (TEKS), which are the teaching standards for the state of Texas. The TEKS, for example, devote a significant portion of the objectives to the Nature of Science and process skills and the rest to content knowledge.

The teaching of science as inquiry embedded in the *teach*HOUSTON program is also recommended by a myriad of resources including the *National Science Education Standards* (National Research Council [NRC], 1996); *Benchmarks for Science Literacy* (American Association for the Advancement of Science, 1993); *Rising above the Gathering Storm: Energizing and Employing America for a Brighter Economic Future* (National Academy of Sciences, 2007); and *America's Lab Report: Investigations in High School Science* (NRC, 2005). The National Science Education Standards (NRC, 1996), for instance, defines inquiry as follows: Scientific inquiry refers to the diverse ways in which scientists study the natural world and propose explanations based on the evidence derived from their work. Inquiry also refers to the activities of students in which they develop knowledge and understanding of scientific ideas, as well as understandings of how scientists study the natural world (NRC, 1996, p. 23).

The features of classroom inquiry and their variations as endorsed by the NRC are outlined in Table 1:

Despite the emphasis placed on the use of inquiry in science education being addressed in preservice and in-service science teacher preparation programs, research has shown that the majority of teachers still fail to incorporate inquiry teaching methods into their teaching repertoires (Goodrum & Rennie, 2007; MacLeod, 2013; Silm, Tiitsaar, Pedaste, Zacharia, & &Papaevripidou, 2017). *teach*HOUSTON addresses this need not only to include inquiry in science teacher education but also to promote continued use of inquiry within the Texas educational landscape.

THE *teach*HOUSTON PROGRAM AND ITS EVOLUTION

I teach to give something back that was given to me.

–Jeff Morgan, Co-Founder of *teach*HOUSTON

Jeff Morgan, now Associate Provost of Education Innovation and Technology, Director of the Center for Academic Support and Assessment (CASA), and Co-director for the *teach*HOUSTON program, was instrumental in creating the *teach*HOUSTON program at the University of Houston. He is a huge proponent of STEM teacher preparation and firmly believes that teacher preparation lies within the responsibilities of Universities. In his words, "At that time, universities were failing. We were not producing enough people with degrees in STEM areas that were going into teaching." Additionally, the university was producing very few high school mathematics teachers, and there was a huge surge in alternative certification in Houston. "People were getting certified to teach mathematics and science who had little training in the STEM areas." Thus, when the founder of UTeach at the University of Texas at Austin met with leaders in the Colleges of Education and Natural Sciences and Mathematics, his interest was piqued. At the time, Jeff Morgan was the chair of the Department of

Table 1. Essential Features of Classroom Inquiry and Their Variations

Essential Feature	Variations			
1. Learner engages in scientifically oriented questions	Learner poses a question	Learner selects among questions, poses new questions	Learner sharpens or clarifies question provided by teacher, materials, or other resources	Learner engages in questions provided by teacher, materials, or other sources
2. Learner gives priority to **evidence** in responding to questions	Learner determines what constitutes evidence and collects it	Learner directed to collect certain data	Learner given data and asked to analyze	Learner given data and told how to analyze
3. Learner formulates **explanations** from evidence	Learner formulates explanation after summarizing evidence	Learner guided in process of formulating explanations from evidence	Learner given possible ways to use evidence to formulate explanation	Learner provided with evidence and how to use evidence to formulate explanation
4. Learner connects explanations to scientific knowledge	Learner examines other resources and forms the links to explanations	Learner directed toward areas and sources of scientific knowledge	Learner given possible connections	
5. Learner communicates and justifies explanations	Learner forms reasonable and logical argument to communicate explanations	Learner coached in development of communication	Learner provided broad guidelines to use sharpen communication	Learner given steps and procedures for communication
More	——— **Amount of Learner Self-Direction**	———	———	**Less**
Less	——— **Amount of Direction from the Teacher or Material**	———	———	**More**

Source: (NRC, 2000, p. 23).

Mathematics in the College of Natural Sciences and Mathematics (NSM), and he embarked on a journey with Dean Robert Wimpelberg of the College of Education (COE) to replicate the UTeach program at the University of Houston. UTeach, a nationally acclaimed STEM teacher preparation program, began in 1997 in order to recruit STEM majors into teaching and to combat the shortage of qualified mathematics and science teachers. *teach*HOUSTON became the first of 45 UTeach replication sites. Integral to the successful replication of the UTeach program at UH was an effective collaboration between the STEM and education colleges which required breaking down discipline silos and commitment from stakeholders in both colleges. Morgan and Wimpelberg worked together to create curricula for the program that included both STEM and education content and pedagogy. This was important as research suggests that successful interdisciplinary collaborations such as those between the colleges of science and education may benefit teacher preparation programs (Evans, Dillard,

Rodriguez-Wilhelm, & McAlister-Shields, 2019). Therefore, the leaders from both colleges acted as boundary spanners (Joshi, Pandey, & Han, 2009) in order to create pathways for the success of preservice STEM teachers. Thus, the *teach*HOUSTON STEM teacher preparation program was birthed in 2007 to address the critical need for highly qualified STEM teachers in Texas and across the country (National Academy of Sciences, 2007).

*teach*HOUSTON started with 14 students and to date has approximately 370 students. The program has goals which are in alignment with working to meet the goal of the *Prepare and Inspire* report to train 100,000 new quality STEM middle and high school teachers (President's Council of Advisors on Science and Technology, 2010). *teach*HOUSTON provides compact and flexible degree plans that fully integrate grade 7–12 teacher certification for those obtaining a major in STEM without adding time or cost to four-year degrees. STEM teachers are prepared through early and ongoing field experiences and rigorous research-based instruction integrating content and pedagogical knowledge provided by faculty members that have extensive teaching experience in public schools. *teach*HOUSTON produces diverse teachers with solid content knowledge for enhancing student learning, which will increase the number of students from underrepresented groups entering into STEM-related majors and professions. From the beginning to the end of their teacher education program and beyond, *teach*HOUSTON graduates are charged with changing the face of public education.

Paramount to the success of the *teach*HOUSTON program are the essential pillars of success including (1) a collaboration between NSM and COE; (2) active recruitment of NSM majors to take the two initial one-hour *teach*-HOUSTON courses to "try out" teaching as an alternate career; (3) a focus on developing deep-level understanding of the subject material and incorporating effective approaches using technology in teaching; (4) guidance and inspiration provided by faculty and highly experienced public school teachers who serve as master teachers; (5) compact degree plans allowing students to complete their major and *teach*HOUSTON courses in four years; (6) courses taught by faculty members experienced in teaching in public schools and actively engaged in STEM education research; (7) integrated research-based professional development courses that focus on teaching both mathematics and science, and (8) a wide array of informal student experiences and leadership opportunities through outreach efforts around the Greater Houston area (Craig et al., 2017).

The *teach*HOUSTON program combines undergraduate degrees in science, computer science, or mathematics with learning how to teach (pedagogy), field-based training (practicums), and teacher certification. As previously mentioned, key components of the program include the following:

- *Strong collaboration* between the College of Natural Sciences and Mathematics and the College of Education is necessary to ensure the highest quality preparation in both subject matter and teaching skills. Preservice teachers receive their major in mathematics or science and their capstone minor along with their teacher certification.

- *Active recruitment* of science and mathematics majors early in their academic careers to take the two initial *teach*HOUSTON courses, Step 1 and Step 2. Students are immediately afforded the opportunity to try out teaching in public schools where they are placed with mentor teachers who are experts in their craft.
- *Early and intensive field experiences* throughout the program, including the opportunity for students to teach lessons as early as their first semester, become increasingly comfortable with teaching in a variety of school environments as they advance through the program. Preservice teachers begin teaching upper elementary students in their first field-based course and work with middle school students in their second field-based course. The third and fourth courses have a high school field experience followed by a semester long experience of student teaching which is typically at the high school level.
- *Master teachers* with extensive teaching and leadership experience in the public schools work at the University of Houston alongside students in the program. The master teachers teach and coordinate field experiences associated with the *teach*HOUSTON program. They work with students as they progress through the program and graduate and then continue to work with them as new teachers for up to three years after they start full-time employment.
- *Mentor teachers*, who are currently teaching in public schools and are highly recommended by their administrators, work closely with *teach*HOUSTON students during their field experiences. Additionally, mentor teachers work with preservice teachers to develop and teach two or three inquiry-based lessons during the semester for the four field-based courses that occur prior to the entire semester of student teaching. Constructive feedback is provided by the mentor teachers subsequent to each teaching experience.
- *Faculty members* at UH are actively engaged in research in mathematics and science, as well as the teaching and learning of mathematics and science. They focus on developing deep-level understanding of the subject area material and incorporating effective teaching approaches with a strong emphasis on the integration of technology.
- *Appealing student opportunities* which promote enrollment and retention, including paid internships that offer education-oriented community outreach including summer STEM camps as well as scholarships limited to those in the *teach*HOUSTON program. For example, the National Science Foundation Robert Noyce Scholarship and Internship program at UH offers both internships and scholarships for preservice teachers. The Noyce Scholarship program offers junior and senior level STEM majors seeking secondary teacher certification $12,000/year scholarships. Students are eligible to apply for the scholarship for up to two years. All students accepting a Noyce scholarship are required to sign a contract that requires them to teach for two years in a high-need school district for every year of scholarship support after graduation.

- *Informal experiences* to better prepare teachers to be effective in high-need schools and provide opportunities for students to build skills and increase self-efficacy as educators. Opportunities include the National Science Foundation Noyce Internship Institute to introduce students to teaching early in their academic careers, multiple professional development experiences, community service events, and networking sessions.
- *Compact, time-sensitive degree plans* that allow most students to graduate within four years with a bachelor's degree in mathematics or science *and* teacher certification. Two tracts, a bachelor of science (BS) and a bachelor of arts (BA) degree with a *teach*HOUSTON options, have been developed for biology, chemistry, physics, mathematics, and computer science majors. The BS degree programs include the prescribed courses for a typical mathematics or science degree with many of the University required core courses scheduled to be taken over the summer sessions. Typically, a BA degree is pursued by students who intend to teach in the 7–12 arena; therefore, the designed BA degree plan with a *teach*HOUSTON option is a viable alternative for students entering the program. It offers a major course plan that includes courses which would be more beneficial for teaching versus the courses included in the BS degree with a *teach*HOUSTON option that are geared more for students moving into technical STEM careers or graduate programs. It includes the major courses necessary for adequate preparation of mathematics and science teachers; however, it eliminates some of the advanced specialty major courses that are not required for teaching at the 7-12 level. Recently, the College of Technology, with *teach*HOUSTON, devised degree plans for technology majors seeking teacher certification. The first cohort of these majors will begin in the fall 2020 semester through funding from a newly awarded National Science Foundation Grant.

REQUIRED COURSEWORK FOR *teach*HOUSTON CERTIFICATION

The nine required teacher certification courses for the *teach*HOUSTON program are aligned with the Texas Education Agency's Approved Educator Standards (Texas Education Agency, 2019). These include Step 1: Inquiry Approaches to Teaching; Step 2: Inquiry-Based Lesson Design; Knowing & Learning in Mathematics and Science; Classroom Interactions (CI); Perspectives in Science and Mathematics; Research Methods; Multiple Teaching Strategies (MTS); Student Teaching Seminar; and Student Teaching (ST). These courses are in the department education with the exception of Research Methods which is listed as a biology (BIO), chemistry (CHEM), or physics (PHYS) course, depending on the student's major. The curriculum and instruction courses have the prefix CUIN and the student teaching courses have the prefix EDUC. The courses integrate the concepts of mathematics and science with a technology-based curriculum which utilizes interactive, inquiry-based teaching tools. Inquiry-based teaching and learning have become key components of science education in recent years (American Association for the Advancement of Science, 1993; National Research Council, 2005) and a foundational part of teacher preparation in *teach*HOUSTON. Program courses cover (1) methods of teaching science and mathematics in schools; (2) a variety of theories and frameworks addressing how people learn mathematics and science; (3) multiple models of teaching, what each model requires of teachers, and the corresponding impact on interactions that occur in mathematics and science classrooms. The Step 1 and Step 2 courses, CI, MTS, and ST, incorporate field experiences which provide students the opportunity to implement instructional strategies and experience success and failure in a highly supportive setting. Students enrolled in field-based courses work with mentor teachers in local area schools. The remaining courses are focused on development and application of pedagogical content knowledge. Another course, Physics by Inquiry, was created in 2013 and addresses both physics content and instruction and counts as an upper level physics/science course for the BA/BS physics degree plans and/or may substitute for the perspectives course for non-physics majors/minors.

Table 2 provides an outline of the courses needed for certification followed by a description of the field-based courses. Additionally, students take required workshops/trainings in dyslexia, code of ethics, mental health, substance abuse and youth suicide, classroom management, teacher/principal evaluation, appropriate relationships, digital learning, and reading instruction. Through the *teach*HOUSTON program, preservice teachers receive over 500 hours of field-based teaching experiences prior to graduation.

Table 2. *Description of teach*HOUSTON *Courses Needed for Teaching Certification.*

Course	Title	Type	HRS
CUIN 1101	Teaching Science and Math-STEP 1	Field-based; Recruitment Course	1
CUIN 1102	Teaching Science and Math-STEP 2	Field-based; Recruitment Course	1
CUIN 3350	Knowing and Learning *Science* and Math	STEM Education Course	3
CUIN 3351	Classroom Interactions in Science and Math	Field-based; STEM Education Course	3
CUIN 3352	Perspectives on Science and Mathematics	STEM Education Course	3
PHYS 4342	Physics by Inquiry *Replaces CUIN 3352	Specialized STEM Content Course **developed with NSF grant 1240083**	3
CUIN 4350	Multiple Teaching Strategies in Science and Math	Field-based; STEM Education Course	3
CHEM, BIO, PHYS 4340	Research Methods in Science and Math	Specialized STEM Content Course **(modified to inquiry-based with NSF grant 1557309)**	3
EDUC 4314 and 4315	Student Teaching and Student Teaching Seminar	Field-based; STEM Education Course	6

Course Descriptions

In Step 1, students explore teaching science or mathematics as a career. In this one credit hour course, students teach science or mathematics lessons in elementary classrooms to obtain firsthand experience with implementing inquiry-based curriculum. Master teachers – clinical faculty employed by the university with extensive, exemplary K-12 teaching experience – teach the courses giving students a first taste of teaching in a diverse yet supportive environment. In this course, students focus on developing their understanding of effective components of inquiry-based lesson plans and delivering highly effective lessons to elementary students, as well as thoughtful reflection on their classroom observations and teaching experiences. Key instructional approaches include classroom discussion, lesson demonstration, student reflection, collaboration, and peer coaching. Demonstration lessons, led by the course instructors, are carefully selected to serve a number of purposes. They not only model promising instructional practices but also address relevant state science or mathematics standards.

Step 2 invites students who want to continue exploring teaching as a career to become familiar with the middle school environment by observing, experiencing, and reflecting on middle school culture and by developing and teaching several lessons based on district curriculum to a middle school class. In Step 2, students observe exemplary teachers and obtain field experience planning and teaching two mathematics or science lessons in local middle schools with diverse student populations. Students in this course receive mentoring from high-quality classroom teachers, as well as instructor feedback based on field observations. The course emphasizes a research-based instructional lesson plan model, aligning lessons to district curriculum, attributes of adolescent students, utilization of technology by teachers and students, and formal and informal methods of assessment. Key instructional approaches include classroom discussion, lesson demonstrations, student reflection, collaboration, and peer coaching.

In Classroom Interactions, students participate in several learning activities which allow them to evaluate their own learning and understanding of a topic. Participating in learning activities also allows students to consider equity issues. For example, is it fair for only the fastest students to contribute to an activity? How would learning be different if all students were not only allowed but required to participate? Is it fair that some students are learning in a language that is not their first? The class considers the implications of deficit thinking in classroom outcomes. Students interview and observe classroom teachers and teach in high school classrooms. The first teaching experience is a 1-day event; the second lasts 2 days. Both teaching experiences are videotaped. Students spend significant time preparing, practicing, and revising lessons for the teaching events. Master and mentor teachers work closely with students on lesson preparation and implementation. They assist with the development of sample activities linked to the content to be taught during the field experience. Key instructional approaches include intensive coaching of students in developing multiday inquiry-based lessons; modeling of effective instructional strategies that illustrate equitable treatment of students; demonstration of effective questioning strategies; implementing cooperative learning groups; interactive classroom discussions about research on teaching and learning and application to classrooms; student presentations that demonstrate analysis and reflection of their teaching; and peer collaboration and coaching.

With the Multiple Teaching Strategies class, students experience and analyze a variety of teaching strategies. Model lessons involving several of the strategies are presented to (1) demonstrate appropriate usage of the strategy and (2) allow students to experience effective implementation of the strategy. Some lessons are presented twice using a different teaching strategy each time. Attributes (strong and weak) of each strategy are addressed, thus allowing for discussion of whether one strategy is more effective for the given topic than another. This leads to discussions of how to choose the most effective teaching strategy for the given material. Teaching strategies that are less familiar to most learners (i.e., inquiry, discovery, problem solving, and project-based approaches to teaching) are the focus of the course. Students observe exemplary high school teachers as well as develop and teach lessons in diverse, urban high schools. Students also develop and present a project-based instructional unit. Teaching strategies that enhance all models of teaching are modeled and incorporated into lessons taught in area high schools.

Student teaching is a full-time immersion into emerging urban and urban public school settings that involves a 14-week, all-day experience and provides *teach*HOUSTON students with a culminating experience that is sufficiently rich and deep in the expectations, processes, and rewards of teaching. Because *teach*HOUSTON students engage in numerous field experiences prior to the student teaching, it is expected that *teach*HOUSTON student teachers assume a teaching role very quickly in the semester and will begin to explore leadership opportunities beyond the classroom (e.g., attending conferences, publications, professional memberships, professional presentations, etc.). Preservice teachers work with University Supervisors and Cooperating Teachers to achieve course objectives, and activities are aligned with the State Board for Educator Certification's Learner Centered Proficiencies, the standards for all new teachers in Texas. Upon completing this course, student teachers will be able to design instruction appropriate for all students that reflects an understanding of relevant content and is based on continuous and appropriate assessments; create a classroom environment of respect and rapport that fosters a positive climate for learning, equity, and excellence; promote student learning by providing responsive instruction that makes use of effective communication techniques, instructional strategies that actively engage students in the learning process, and timely high-quality feedback; and fulfill professional roles and responsibilities and adhere to legal and ethical requirements of the profession.

CULTURALLY RESPONSIVE PEDAGOGY

Despite the 'stated desire of the United States to educate all youth, children of color attending urban schools struggle to succeed (Howard & Milner, 2014). Given any predictive measure of success, such as standardized test scores, college and high school grade point averages (GPA), graduation and dropout rates, children of color, particularly Hispanic American and African American students, do not achieve academically at the same rate as their European American counterparts (Banks & Banks, 2013; Blanchett, 2014). Furthermore, African American students have been placed in special education programs and/or are entered into disciplinary programs at higher levels than their European American counterparts (Gregory & Mosley, 2004; Tefera, Thorius, & Artiles, 2014). This academic gap, alternately termed the achievement gap or the opportunity gap (Darling-Hammond, 2011), exists regardless of socioeconomic status, gender, or national geographic location (Haycock, 2001; Ladson-Billings, 2014). When examining the research to improve the education of children of color, the preponderance of evidence suggests that teacher attitudes and specific instructional behaviors have an impact on the academic achievement of underrepresented youth (Gay, 2014; Irvine, 2003). Darling Hammond's metaanalysis of policy documents from 50 states maintains that teacher effects are "additive and cumulative" (Darling-Hammond, 2000). International research (Fullan, 2011; McKinsey & Co, 2007; OECD, 2010) furthermore confirms that "the single most significant means of improving education… is through teaching" (Pollard, 2010, p. 4). In the words of the UK Department for Education, "the first, and most important, lesson is that no education system can do better than the quality of its teachers" (Department for Education, 2010, p. 3).

Although teacher education programs and professional development for practicing teachers provide preparation for teaching children from historically underrepresented cultures, Gay (2010) states that the application of that knowledge in the classroom is often inconsistent and ineffective because a theory–practice disconnect frequently occurs. The missing component is development in culturally responsive pedagogy (CRP) (Howell, Lewis, & Carter, 2011; Marrongelle, Paola, & Smith, 2013) paired with on-site training with teachers (Gay, 2010; Green & Allen, 2015; Hirsh, 2005). These corrective actions have proven to be successful in raising students' academic achievement and attainment and transforming classrooms populated by underrepresented youths in urban schools such as those in Greater Houston. High-quality preparation in the STEM disciplines is important, but unfortunately not enough to ensure success in the urban area's public schools. Additional complexities of the urban school situation need to be addressed. CRP is a major concept in urban and multicultural education that stresses the ability of teachers to respond to their students by incorporating elements of students' culture in their teaching (Gay, 2014). According to Gay (2010), CRP uses cultural knowledge, prior experience, frames of reference, and performance styles of underrepresented students to make learning encounters more relevant (see Fig. 3). Research confirms that CRP is a vehicle for increasing student success in high-needs classrooms (Howell et al., 2011; Perry & Delpit, 1998). As a result, *teach*HOUSTON recently incorporated a CRP strand throughout the field-based courses as part of *teach*HOUSTON's evolving trajectory as one of the leading secondary STEM teacher education programs in Texas and in the nation. This CRP strand is

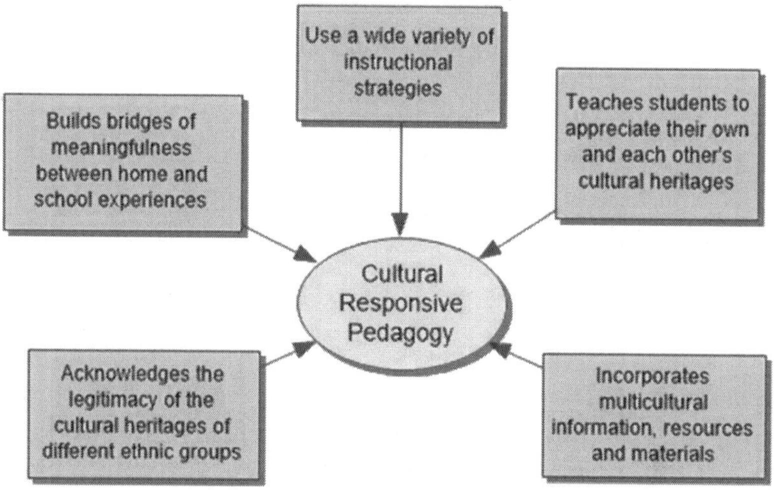

Fig. 3. Characteristics of Culturally Responsive Pedagogy.

vertically aligned with the coursework, scaffolding preservice teachers' understanding and ability to implement culturally responsive instruction. Table 3 gives details on how CRP activities have been incorporated into the aforementioned field-based courses.

CONCLUSION

Since 2007, *teach*HOUSTON has prepared an increasing number of secondary science teachers with urban teaching experiences who mostly will teach in metropolitan Houston schools. The following percentages characterize *teach*-HOUSTON graduates: 88% of the program's graduates are teaching in the Greater Houston area; 80% of the graduates are teaching in high-need schools; 87% of the graduates continue as public school teachers beyond five years; and the program graduates one of the largest pools of chemistry teachers in the country (top 2% of universities nationally). Moreover, *teach*HOUSTON's graduates reflect UH student diversity (Fig. 4) with 34% Hispanic, 30% Asian, 25% White, 8% African American, and 3% other as compared to US and Texas teachers (Boser, 2014), which is essential as successful academic outcomes occur when minority teachers serve as role models (Dee, 2005; McNamara & Basit, 2004; Pitts, 2007).

As the number of teachers produced by the program increases, the number of largely underserved minority, middle, and high school students who have been or will be taught by qualified *teach*HOUSTON teachers with strong science content area knowledge and pedagogy will also increase as is shown in Fig. 5.

Table 3. Incorporation of Culturally Responsive Pedagogy (CPR) into *teach*HOUSTON Field-based Courses.

Class	Class Description	CRP Activities	CRP Description
CUIN 1101	Step 1: Teaching Science and Mathematics	Model in 5E Lesson Plans	Students experience lessons inclusive of CRP.
CUIN 1102	Step 2: Teaching Science and Mathematics	What is Culture? Model in 5E Lesson Plans; Introduction of CRM/ST Lesson Plan Analysis	Students begin the reflexive process by engaging in the culture activity to learn the various aspects of culture and how culture informs instruction.
CUIN 3351	Classroom Interactions	Funds of Knowledge; The Tale of Two Nora's; Utilization of CRM/ST Lesson Plan Analysis	Students explore more deeply their own funds of knowledge and how culture can be used as a cognitive scaffold by analyzing lessons plans for CRP and also distinguishing between two different styles of classrooms, one inclusive of CRP and one not inclusive of CRP.
CUIN 4350	Multiple Teaching Strategies	Chalk Talk; Video and Lesson Plan Analysis; Implementation in Project Based Learning	Students delve deeper into CRP with a Chalk Talk Activity and the analysis of lesson plans. Students implement CRP in their Project-Based Learning Unit.
CUIN 4314 and 4315	Student Teaching and Student Teaching Seminar	Integration and Reflection; Observe CRP Lessons; Community Walk/Culture Audit	Students further integrate CRP into their lesson plans, observe and analyze other lessons inclusive of CRP, and engage in a community walk.

Thus far, the *teach*HOUSTON program has successfully enabled the University of Houston to contribute significantly to the national effort to improve mathematics and science education. As explained throughout this chapter, it has become increasingly difficult in the United States and internationally to recruit, train, and develop teachers with a long-term commitment to teaching, especially in the areas of mathematics and science content. Today, 80% of *teach*HOUSTON's 347 graduates work in the local, urban area with predominantly at-risk, minority students. To conclude, we have distilled what has been learned about *teach*HOUSTON and its challenges and successes into a model of a science as inquiry teacher education program (Fig. 6) that we hope will be of value to others addressing similar problems in the field and in the teaching of science internationally. As the figure shows, *teach*HOUSTON's next steps involve the fuller development of graduates' induction year (first 3–5 years of career) experiences, which presumably will increase teacher retention in the urban area over time.

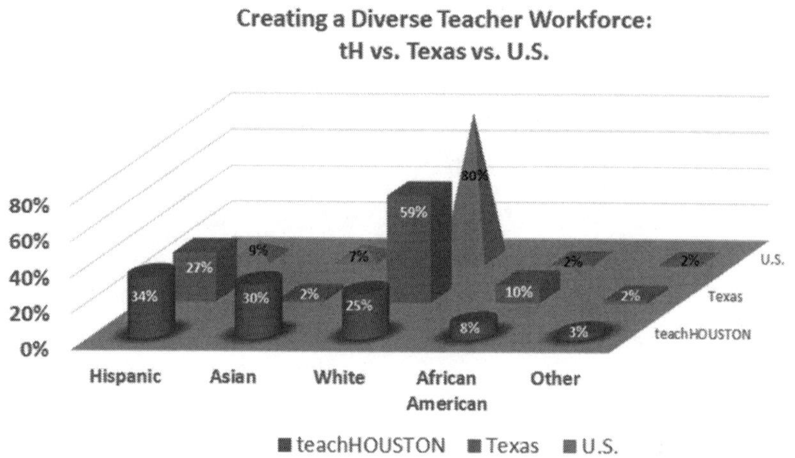

Fig. 4. Ethnicity of *teach*HOUSTON Teachers vs US vs Texas teachers.

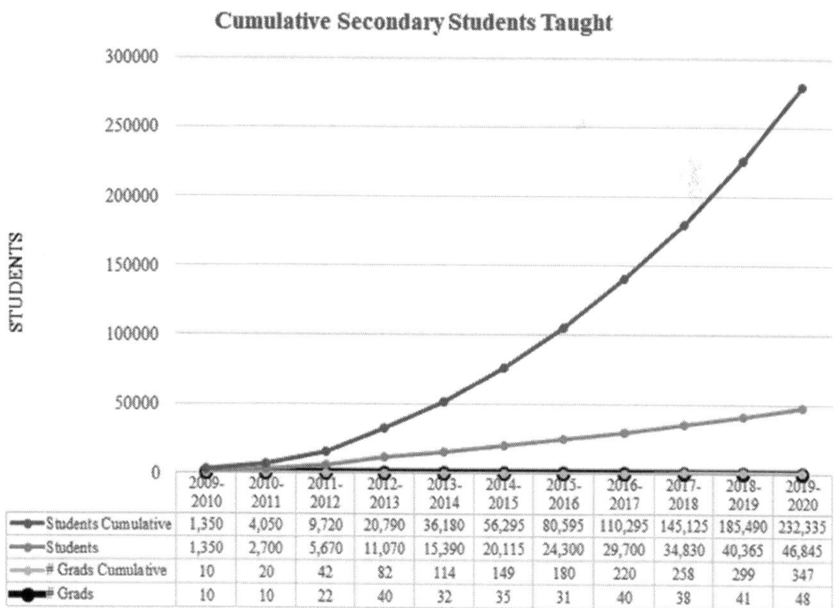

	2009-2010	2010-2011	2011-2012	2012-2013	2013-2014	2014-2015	2015-2016	2016-2017	2017-2018	2018-2019	2019-2020
Students Cumulative	1,350	4,050	9,720	20,790	36,180	56,295	80,595	110,295	145,125	185,490	232,335
Students	1,350	2,700	5,670	11,070	15,390	20,115	24,300	29,700	34,830	40,365	46,845
# Grads Cumulative	10	20	42	82	114	149	180	220	258	299	347
# Grads	10	10	22	40	32	35	31	40	38	41	48

Fig. 5. Cumulative Secondary Science Students Taught by *teach*HOUSTON **Graduates** (Cumulative Number of Students Taught Is an Estimate that Assumes that 90% of Graduates Will Go into Teaching and Will Teach 150 Students per Year.)

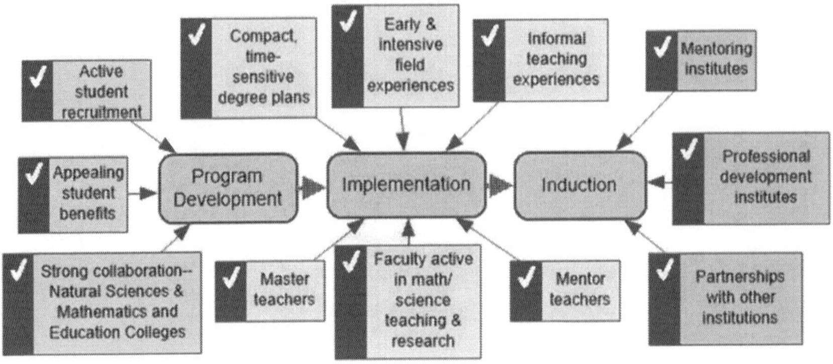

Fig. 6. Model of Science as Inquiry Teacher Education Program.

AUTHORS' NOTE

This research was funded by the following grants:

- National Science Foundation DUE Award 1240083, *Recruitment, Preparation and Retention of STEM Students as High School Teacher*
- National Science Foundation DUE Award 1557309, *University of Houston: Learning through Informal and Formal Experiences*
- National Science Foundation DUE Award 1557273, *Collaborative Research: Understanding Robert Noyce Teacher Scholarship Outcomes in Texas*
- National Science Foundation DUE Award 1759454, *Leadership through Equity and Advocacy Development* (LEAD Houston)
- Department of Education Supporting Effective Educator Development (SEED) Grant, U423A180139, *Expanding and Strengthening the STEM Teacher Workforce Through UTeach*

A special thank you is extended to Dr. Liping Wei who also contributed to this project. We additionally acknowledge the contributions of Dr. Gayle Curtis who assisted Paige Evans in the creation of this chapter's figures.

REFERENCES

American Association for the Advancement of Science. (1993). *Benchmarks for science literacy: A project 2061 report*. New York, NY: Oxford University Press.

Banks, J. A., & Banks, C. A. (2013). *Multicultural education: Issues and perspectives* (8th ed.). Hoboken, NJ: John Wiley & Sons.

Blanchett, W. (2014). African American students and other students of color in special education. In H. Milner, & K. Lomotey (Eds.), *Handbook of urban education* (pp. 271–284). New York, NY: Routledge.

Boser, U. (2014). *Teacher diversity revisited, a new state-by-state analysis*. Washington, DC: Center for American Progress.

Chen, C. S. (2009). *Numbers and types of public elementary and secondary local education agencies from the common core of data: School year 2007-08. First look. NCES 2010-306*. Washington, DC: U.S. Department of Education. National Center for Education Statistics.
Chenowith, K. (2010). Turning obstacles into triumph. *US News & World Report*, *147*(7), 58–61.
Craig, C. (2014). From stories of staying to stories of leaving: A U.S. beginning teacher's experience. *Journal of Curriculum Studies*, *46*(1), 81–115.
Darling-Hammond, L. (2000). Teacher quality and student achievement: A review of state policy evidence. *Education Policy Analysis Archives*, *8*, 1.
Darling-Hammond, L. (2011). *The world is flat and the educational future: How America's commitment to equity will decide our future*. New York, NY: Teachers College Press.
Dee, T. S. (2005). A teacher like me: Does race, ethnicity, or gender matter? *The American Economic Review*, *95*(2), 158–165.
Department for Education. (2010, November). *The importance of teaching: The schools white paper*. London: Her Majesty's Stationary Office.
Dewey, J. (1938). *Experience and education*. New York, NY: Basic Books.
Evans, P. K., Dillard, K. C., Rodriguez-Wilhelm, D., & McAlister-Shields, L. (2019). Like-minded people: University-based interdisciplinary collaborations in STEM teacher preparation programs. *Journal for STEM Education Research*, *2*(1), 35–54.
Fullan, M. (2011). *The new meaning of educational change* (4th ed.). New York, NY: Teachers College Press.
Fuller, E. (2009). *Secondary mathematics and science teachers in Texas: Supply, demand, and quality*. Texas Instruments & TBEC.
Gay, G. (2010). Acting on beliefs in teacher education for cultural diversity. *Journal of Teacher Education*, *61*(1–2), 143–152.
Gay, G. (2014). Culturally responsive teaching principles, practices, and effects. In R. Milner IV, & K. Lomotey (Eds.), *Handbook of urban education* (pp. 353–372). New York, NY: Taylor & Francis.
Goodrum, D., & Rennie, L. J. (2007). *Australian school science education national action plan, 2008-2012*. Canberra: Department of Education, Science and Training.
Green, T., & Allen, M. (2015). Professional development urban schools: What do teachers say? *Journal of Inquiry & Action in Education*, *6*(2), 53–79.
Gregory, A., & Mosley, P. (2004). The discipline gap: Teacher's views on the over-representation of African American students in the discipline system. *Equity & Excellence in Education*, *37*, 18–30.
Haycock, K. (2001). Closing the achievement gap. *Educational Leadership*, *58*, 6–11.
Hirsh, S. (2005). Professional development and closing the achievement gap. *Theory into Practice*, *44*, 38–44.
Howard, T., & Milner, H. (2014). Teacher preparation for urban schools. In H. Milner, & K. Lomotey (Eds.), *Handbook of urban education* (pp. 199–216). New York, NY: Routledge.
Howell, L., Lewis, C., & Carter, N. (2011). *Yes we can: Improving urban schools through innovative education reform*. Charlotte, NC: Information Age Publishing.
Irvine, J. J. (2003). *Educating teachers for a diverse society: Seeing with the cultural eye*. New York, NY: Teachers College Press.
Joshi, A., Pandey, N., & Han, G. (2009). Bracketing team boundary spanning: An examination of task-based, team-level, and contextual antecedents. *Journal of Organizational Behavior*, *30*(6), 731–759. Retrieved from http://www.jstor.org/stable/41683864.
Ladson-Billings, G. (2014). Race, research and urban education. In H. Milner, & K. Lomotey (Eds.), *Handbook of urban education* (pp. 433–452). New York, NY: Routledge.
MacLeod, K. (2013). Physics education and STSE: Perspectives from the literature. *European Journal of Physics Education*, *4*(4), 1–12. Retrieved from http://www.eu-journal.org/index.php/EJPE/article/view/96.
Marrongelle, K., Paola, S., & Smith, M. (2013). Scaling up professional development in an era of common state standards. *Journal of Teacher Education*, *64*, 202–211.
McKinsey & Co. (2007). How the world's best performing school systems come out on top. *McKinsey & Co*. Retrieved fromhttp://mckinseyonsociety.com/downloads/reports/Education/Worlds_School_Systems_Final.pdf.

McNamara, O., & Basit, T. N. (2004). Equal opportunities or affirmative action? The induction of minority ethnic teachers. *Journal of Education for Teaching*, *30*(2), 97–115.

National Academy of Sciences. (2007). *Rising above the gathering storm: Energizing and employing America for a brighter economic future*. Washington, DC: National Academy Press.

National Public Radio. (2014, July 18). The teacher dropout crisis. Retrieved from http://www.npr.org/sections/ed/2014/07/18/332343240/the-teacher-dropout-crisis.

National Research Council. (1996). *National science education standards*. Washington, DC: National Academy Press.

National Research Council. (2000). *Inquiry and the national science education standards; A guide for teaching and learning*. Washington, DC: National Academy Press.

National Research Council. (2005). *America's lab report: Investigations in high school science*. Washington, DC: National Academy Press.

Nelson, J., Palonsky, S., & McCarthy, M. R. (2009). *Critical issues in education: Dialogues and dialects* (7th ed.). New York, NY: McGraw-Hill.

OECD. (2010). *Education at a glance 2010: OECD indicators*. Paris: OECD.

Perry, T. & Delpit, L. D. (Eds.). (1998). *The real ebonics debate: Power, language, and the education of African-American children*. New York, NY: Beacon Press.

Pitts, D. W. (2007). Representative bureaucracy, ethnicity, and public schools: Examining the link between representation and performance. *Administration & Society*, *39*(4), 497–526.

Pollard, A. (2010). Professionalism and pedagogy: A contemporary opportunity. A commentary by the Teaching and Learning Research Programme and the General Teaching Council for England.

President's Council of Advisors on Science and Technology. (2010). *Prepare and inspire: K-12 education in science, technology, engineering, and math (STEM) for America's future*. Washington, DC: Executive Office of the White House.

Schwab, J. (1962). The teaching of science as enquiry. In J. Schwab, & P. Brandwein (Eds.), *The teaching of science* (pp. 3–103). Cambridge, MA: Harvard University Press.

Silm, G., Tiitsaar, K., Pedaste, M., Zacharia, Z. C., & Papaevripidou, M. (2017). Teachers' readiness to use inquiry-based learning: An investigation of teachers' sense of efficacy and attitudes toward inquiry-based learning. *Science Education International*, *28*(4), 315–325.

Tefera, A., Thorius, K., & Artiles, A. (2014). Teacher influences in the racialization of disabilities. In H. Milner, & K. Lomotey (Eds.), *Handbook of urban education* (pp. 114–128). New York, NY: Routledge.

Texas Education Agency. (2017). Out of field teaching report, 2016-2017. Retrieved from https://tea.texas.gov/Reports_and_Data/Educator_Data/Educator_Reports_and_Data/.

Texas Education Agency. (2019, August 26). Approved educator standards. Retrieved from https://tea.texas.gov/Texas_Educators/Preparation_and_Continuing_Education/Approved_Educator_Standards.

U.S. Department of Education. (2016). 2016 title II report: National teacher preparation data. Retrieved from https://title2.ed.gov/Public/SecReport.aspx.

Villegas, A. M., & Clewell, B. C. (1998). Increasing the number of teachers of color for urban schools. *Education and Urban Society*, *31*, 42-61.

Xue, Y., & Larson, R. (2015). STEM crisis or STEM surplus? Yes and yes. *Monthly Labor Review*. doi: 10.21916/mlr.2015.14.

COLLABORATION BETWEEN A PHYSICS PROFESSOR AND A PHYSICS TEACHER/TEACHER EDUCATOR

Donna W. Stokes, Paige K. Evans and Cheryl J. Craig

ABSTRACT

*Collaborations between faculty from the teach*HOUSTON *program and physics department have led to an increase in the number of highly qualified physics teachers produced by the University of Houston. Faculty were able to systematically build the physics teacher preparation program through the following endeavors: streamlined degree plans, a physics inquiry course, an internship program, a scholarship program, and induction activities for the first three years of their teaching degrees. This has resulted in preparing approximately three physics teachers annually. Prior to this collaboration, the University had not produced any physics teacher graduates in the previous decade.*

Keywords: Physics teacher education; teacher education; inquiry-based teaching; recruitment; retention; collaboration; *teach*HOUSTON

Generally, studies and research on science, technology, engineering, and mathematics (STEM) education have been carried out by researchers in Colleges of Education at institutions of higher education. In particular, in the case of STEM teacher preparation, many faculty in science and mathematics colleges have viewed research, evaluation, and analysis of learning/teaching in STEM as a topic to be handled by those with backgrounds in education. Traditionally, teacher preparation programs are housed in the College of Education as the programs

Preparing Teachers to Teach the STEM Disciplines in America's Urban Schools
Advances in Research on Teaching, Volume 35, 27–40

ISSN: 1479-3687/doi:10.1108/S1479-368720210000035003

entail curriculum and pedagogical preparation required for teacher certification. However, in recent years, teacher education programs are now being housed in the Colleges of Mathematics and Science. For example, the *teach*HOUSTON program at the University of Houston (UH) is the first replication site of the UTeach program, a nationally acclaimed program that places teacher preparation under the auspices of the Colleges of Natural Sciences and Education. The program's (UTeach, 2020) model for preparing highly qualified mathematics and science teachers has been cited by the National Academy of Sciences (2007) and has been replicated at universities across the country. The UH *teach*HOUSTON program brings together faculty from the College of Natural Sciences and Mathematics (NSM), the College of Education, and master teachers from the field to develop teacher education curriculum for secondary STEM teacher certification based on combining the theory and practices of teaching and the STEM content.

Although the *teach*HOUSTON program existed at the UH, since 2007, a focused collaboration between the program and faculty from the Department of Physics did not occur until 2009. This is mainly due to the prevailing attitude that science education research was best explored by education trained researchers and not within the science discipline itself. In the past 20 years, this attitude has changed as physics education researchers have led the expansion of science education research from education departments into specific science colleges and disciplines in an attempt to more closely align content, pedagogy, and practice.

Physics Education Research (PER) was started in the 1970s, although some believe it started as early as the 1900s with the work of John Dewey and Jean Piaget (Cummings, 2011). Others believe it was not a recognized field of study until the mid-1990s when the PER emerged with the majority of the research, done by physics faculty, focused on improving undergraduate physics instruction. At the K-12 level, research/development of physics instructional materials for students by physicists began as early as the 1950s which is when the National Science Foundation (NSF) was created to promote the progress of science (National Science Foundation, 2020). NSF, with the success of the launch of Sputnik in 1957 (National Science Board, 2000), increased funding for science curriculum development; therefore, through their funding, physicists continued to develop leading edge curriculum which was inquiry based and focused on the use of interactive activities to more fully engage students in the scientific process (2003). The curriculum was developed with the intent that teachers would (1) possess a strong content knowledge base, (2) have basic scientific reasoning skills (Fuller, 2003), and (3) be able to easily utilize the materials to teach science as inquiry to their students by referencing the materials and attending professional development workshops and/or summer institutes (McDermott, 1991). However, implementation and sustainability of the curriculum was not widespread as teachers did not have access to the necessary professional development to teach physics content in this manner (McDermott, Heron, Shaffer, & Stetzer, 2006).

In the late 1960s, Physicist Arnold Arons at the University of Washington developed a preservice elementary teacher education course for physical science and later a high school level course taught as part of the research done by their physics education group. The intent was to prepare teachers to teach science content and to gain the skills needed for engaging in the scientific process (Arons, 1976). The course was originally not a part of the physics department programming as physics faculty did not recognize teacher education as a responsibility of the physics departments, but rather that of the colleges of education. This perspective has been the case for many science and mathematics departments in higher education across the nation; however, since then, many have had a shift in opinion and are now integrating these types of courses into their physics program and are hiring faculty with interests and expertise in science education, developing education-based curriculum and courses, and offering undergraduate and graduate science education degrees.

*teach*HOUSTON/PHYSICS COLLABORATION

Why are collaborations between science/physics and teacher education faculty important for preparing physics teachers? Nationally, approximately 75% of physics teachers lack the content knowledge required to adequately teach a high school physics course. Also, nationally, only 35% of new physics teachers have a degree/minor in physics or physics education (Tesfaye & White, 2012). The *National Academies Rising Above the Gathering Storm* report (National Academy of Sciences, 2007) recognized that qualified teachers in the classroom are a strong predictor of STEM student success. In light of these data, it is imperative that physics and teacher preparation faculty work together to increase the number of qualified physics teachers produced to teach, particularly in high-needs schools.

In this chapter, the significance of a collaboration between a faculty in the Department of Physics and *teach*HOUSTON on production of qualified teachers at the UH will be highlighted. The UH is a US Department of Education-designated, Hispanic-serving and Asian-serving institution and the fourth most diverse university in the United States (US News and World Report, 2019), with 23.7% White, 32.4% Hispanic/Latino, 21.1% Asian, 9.9% Black/African American, 0.1% Hawaiian/Pacific Islander, 0.1% Native American, and 12.7% others in Fall 2019 (UH Institutional Research, 2019). UH, located in the heart of Houston, Texas, the fourth largest city in the nation, enrolls over 46,000 students with more than 80% of its graduates remaining in the Houston area five years after graduation. Unfortunately, the UH had not produced any certified physics teachers in 10 years preceding 2012. The Department of Physics recognized that if the University was to address the current and future demands for qualified physics teachers, it must be committed and involved in teacher education programs and recruitment of students with such interests.

The *teach*HOUSTON program has allowed genuine collaborations to be formed between science/math faculty and education faculty to develop courses and curriculum promoting promising practices of teaching and teaching/learning of STEM content. In 2009, the first collaborative effort of the physics department and *teach*HOUSTON began when we (Stokes, Evans) shared our passion for physics and physics teacher education and discussed submitting a proposal that would establish the Department of Physics as a site for developing a model for training physics teachers. Our passion and dedication to improve the number of physics teachers produced were similar but were driven based on our personal and professional journeys which were very different.

I (Donna W. Stokes – author 1) am a professor in physics. I began my academic career in the academy as an assistant professor with scientific research geared toward understanding the structural, optical, and electrical properties of semiconductor materials for the development of novel detectors and lasers for infrared applications. Although my early scholarship was in condensed matter physics, I became increasingly interested in why/how students learned physics and the factors that influenced their success and persistence toward completing physics/STEM degrees. From the start of my career, I was always interested in physics/STEM student success, and as my career continued, I realized that one of the more critical issues facing STEM education researchers is that less than half of undergraduates pursuing STEM degrees actually complete them and this was indeed true for physics majors at the UH. To ensure the education of a diverse and skilled STEM workforce, I broadened my research focus to include physics/STEM education, which identified factors contributing to low student success rates in STEM majors, particularly underrepresented students. I knew their story, because it is my own. For the past 10 years, student success has been my research focus as I have been dedicated to improving physics/STEM student success and STEM/physics teacher education. In 2009, I met Paige Evans, my coauthor, and we quickly identified the lack of participation of the Department of Physics in preparing highly qualified physics teachers.

Paige Evans (author 2) is a Clinical Professor in the *teach*HOUSTON program at the UH. She began her journey at the UH in 2008 subsequent to teaching high school physics for approximately 18 years. As a former high school teacher who also served as the chair of the science department, she realized the need for physics teachers steeped in promising practices in physics including inquiry-based learning. Early on, it was her goal at the UH to develop physics teachers who were strong in both content and pedagogy. To this end, she sought out collaborations with the physics department. After meeting coauthor Donna W. Stokes, they embarked on their journey to prepare the first physics teachers at the UH in over a decade.

Together, we discussed the need for the Physics programs to make a commitment to develop and enact curriculum and degree plans and to recruit/retain high-quality students interested in teaching physics. This led us to submit a letter of intent to have the UH Physics Department become a PhysTEC site. Founded in 2001, PhysTEC is a partnership between the American Physical Society and the American Association of Physics Teachers, which provides guidance to help colleges/universities develop and/or improve and promote their physics teacher education programs for addressing local and national shortages of qualified physics teachers. With funding from the NSF, PhysTEC has created a network of institutions across the nation for sharing practices in physics teacher education programming. It has served as a hub for research dissemination for physics teacher education. As a PhysTEC site, colleges/universities who did not graduate a significant number of physics teachers are provided with funding to develop and enact projects which can serve as models for teacher education physics programs. We proposed that the UH would be a PhysTEC site that aimed to increase the number of physics majors earning a BS/BA degree with teacher certification through recruitment, retention, and preparation of students using interactive-based teaching tools that have been proven to be effective through PER. In addition, the program would provide mentors from the Department of Physics, *teach*HOUSTON, and the local high school physics teacher community to ensure the success of the students in their first three years of teaching.

Although the letter of intent submitted in 2010 was not selected for full proposal submission, we continued to develop our ideas and visions for moving the collaboration forward for establishing a solid physics teacher education pipeline.

PHYSICS/*teach*HOUSTON DEGREE PLANS

In our initial efforts, we systematically worked to integrate and streamline the physics degree plan to include the *teach*HOUSTON certification course work. This work started with the Undergraduate Studies Committee in the Department of Physics which is composed of 5-7 physics faculty who are responsible for oversight of undergraduate physics academic curriculum. Stokes (author 1), the chair of this committee since 2006, worked with the committee members while Evans (author 2) and *teach*HOUSTON faculty worked to integrate the teacher education courses into the 4-year BS and BA physics degree plans. The degree plans which include 120 hours of required core, math and physics courses, and free electives, did not leave much room for incorporation of the 23 credit hours of teacher education course work. Through collaboration and review of the physics electives and teacher education courses in the program, the physics and *teach*HOUSTON faculty determined which courses could meet the requirements for both programs without losing the rigor and integrity of either program requirements. By allowing courses to meet requirements for both programs, incorporation of the teacher education courses only increased the number of credit hours for obtaining the physics degree with the *teach*HOUSTON option or capstone from 120 to 127 credit hours which could still be completed within 4 years; therefore, no additional time or cost to earning the BS/BA degree and gaining teacher certification would be incurred by the student. This critical agreement required not only collaboration but also trust between the faculty in the departments as this allowed science education courses to serve as physics elective toward the physics degrees.

Building a relationship between the faculty and chair in the Department of Physics and the *teach*HOUSTON master teachers was one of the most critical components of moving the collaboration forward and ensuring quality of courses which would be used to meet the physics degree and teacher certification requirements. Physics faculty were concerned that science education courses would not have the same rigor as a course taught by physics faculty with expertise in the physics content. They also questioned whether the course instructors would present physics content in formal and informal learning settings which would not lead to deep conceptual and quantitative learning. Because of Evans' (author 2's) background (i.e., degree in physics, 18 years teaching physics at the high school level, working with one physics faculty on AP exams developing curricula for an inquiry style physics course), the physics faculty were more open to accepting the courses, taught by *teach*HOUSTON faculty, to meet physics degree requirements as long as she was overseeing the curriculum development and assessment and how it would be enacted. The first course to be approved by the physics Undergraduate Studies Committee was the Research Methods (Phys 4340) course, which would be used in place of three credit hours of Advanced Physics electives for the BS/BA physics degree and would be used to meet three credit hours of a teacher education course. One caveat to using this course toward the Physics degrees is that this would only be possible for students who were enrolled in the *teach*HOUSTON program as their capstone requirement.

NATIONAL SCIENCE FOUNDATION ROBERT NOYCE PROGRAM

In 2012, we continued our collaboration and applied for and were awarded an NSF Robert Noyce scholarship grant program that focused on recruitment, training, and retention of science majors to the *teach*HOUSTON program. The scholarship program, which was tailored toward physics and chemistry majors/minors (but not limited to), focused on increasing the pool of qualified STEM teachers by providing (1) a scholarship; (2) a summer camp internship; (3) a Physics by Inquiry course; (4) professional development; and (5) mentoring and cohort building activities. Research for the program focused on understanding how recruitment, retention, and education of the physics teachers through the program components would impact the number of qualified physics teachers produced at the UH.

RECRUITMENT TO THE PHYSICS/*teach*HOUSTON AND NOYCE PROGRAMS

Recruitment and retention of physics majors was key for improving the physics teacher pipeline. Both physics and *teach*HOUSTON faculty agreed that routine recruitment activities were needed and that collaborative efforts would be required to ensure communication of information for each program was clear. Recruitment efforts were established in the first semester General and University physics courses targeting existing Physics majors, high-performing students, and current *teach*HOUSTON students studying non-physics STEM disciplines. Physics faculty for these introductory physics courses invited *teach*HOUSTON student ambassadors and staff to give a brief presentation about the program and its benefits. Degree plans were advertised as a way to demonstrate how physics and other NSM students could complete their respective degrees and teacher certification within 4 years. For example, a physics minor which included *teach*HOUSTON courses was created and this led to the creation of degree plans such as the BS degree in Mathematics with a physics minor. Specialized degree plans such as this one were approved and promoted by the Physics Department and *teach*HOUSTON with the idea that this would help retain students to complete the physics and teacher certification program and gain certification which would allow them to teach physics. Also, through the Noyce Scholarship program, scholarships and internships were used to encourage physics and other NSM majors enrolled in the introductory physics courses to pursue a physics or chemistry major/minor while pursuing teacher certification through *teach*HOUSTON.

NOYCE SCHOLARSHIP AND SUMMER INTERNSHIP

The Noyce scholarships and internships were additionally used as recruitment tools for physics and *teach*HOUSTON. Noyce Scholarships were offered to junior and senior NSM majors and minors and postbaccalaureate students seeking teacher certification in physics or chemistry. Eligible students could apply for scholarships of $12,000 per year for up to two years of support. For each year of acceptance of the scholarship, scholars were required to commit two years of service in a high-need school district after graduation. In addition, a summer internship program was created to introduce students to teaching early on in their academic careers. The Noyce internship program offered support to freshmen and sophomore NSM undergraduate majors for a 6-week summer internship. Noyce interns were provided with the opportunity to attend a two-week professional development institute and to work as camp counselors alongside science master teachers in a four-week middle school STEM summer camp. This included one of the following: (1) ExxonMobil Bernard Harris Summer Science Camp, a residential camp for underserved students entering 6th, 7th, or 8th graders interested in STEM careers, (2) Chiyoda Young Inventors' Academy for students entering 7th, 8th, and 9th grade who are interested in engineering, design and innovation, or (3) StatOil STEM Camp where 5th 8th graders engage in themed integrate STEM activities. The Noyce interns were required to plan physics/chemistry lessons, activities, experiments, and projects for the campers before the start of the camp and to assist master teachers in teaching the lessons during camp operation. Upon completion of the camp, the interns reviewed and improved the science modules and made suggestions for modules to use for future camps.

PROFESSIONAL DEVELOPMENT INTERNSHIP INSTITUTE

While the Noyce scholarship provided financial support for students who were committed to the *teach*HOUSTON program, and the Noyce internship offered hands-on experience to current and perspective *teach*HOUSTON students early in their career, scholars and interns needed to be exposed to professional skills which would allow them to succeed in the teaching profession. Preparing qualified physics teachers consists of not only educating them to teach the physics content but also equipping them with the skills to address student learning challenges, instructional design issues, and classroom management concerns. As part of the Noyce Internship Program, interns and Noyce scholars attended a two-week professional development Internship Institute which consisted of sessions/workshops developed by former interns/scholars, master teachers, and physics faculty. The Internship Institute not only prepared interns/scholars to serve as counselors and teaching assistants for the STEM summer camps but also allowed them to gain skills needed to teach in STEM, particularly physics, classrooms in high-needs schools. The professional development sessions, facilitated by *teach*HOUSTON faculty and master teachers, cover topics including team building, classroom management, professionalism, engineering challenges, conflict resolution, bully prevention, and growth and fixed mindset. Training Guides (see Chapter 6) for the two-week institute were created to outline the weekly session topics/schedules with links to resources and are available for use by other teacher education programs.

PHYSICS BY INQUIRY COURSE

Content training of qualified physics teachers is where the greatest collaboration between physics and teacher education faculty occurred since both are responsible for ensuring the preservice teachers have a strong physics content knowledge and possess the pedagogy to present the material to their students. This collaborative effort was definitely needed to develop and implement a physics teacher education course which would augment the physics knowledge/understanding attained through the introductory General and University Physics course sequences. The Physics Department Undergraduate Studies Committee wanted to be confident that the courses' physics content was rigorous and was presented in a way which would allow the students to best grasp the concepts so they charged Stokes (author 1) with working closely with Evans (author 2) to review course content to ensure rigor needed to meet the physics degree requirements. As part of the Noyce Scholarship Program, the Physics by Inquiry course (Phys 4342) was developed. The course specifically targets physics instruction with the incorporation of inquiry-based content. The curriculum takes into consideration the College and Career Readiness Standards (CCRS) prescribed by the Texas Higher Education Coordinating Board (the education governing board for the state of Texas), which outlined the knowledge required for students to succeed in entry-level college courses. The course material also covers topics listed by the Texas Education Agency's Texas Education Knowledge Skills (TEKS) curriculum, which is the official K-12 curriculum for the state of Texas. The TEKS details the requirements for every course where state-mandated standardized tests measure acquisition of specific knowledge and skills (Texas Education Agency, 2017). Specific topics were included in the course curriculum covering areas where the TEKS were found to weakly correlate with the CCRS in physics.

This freshly minted course is taught as a set of laboratory-based modules that provide an inquiry approach which allows students, in small groups, to develop a deep understanding of physics concepts through hands-on inquiry style experiments. The students use these concepts to develop rules and patterns and determine equations to utilize and solve problems. Peer instruction, making predictions, and analyzing results are essential components of the course; thereby, students will form mental models for understanding the physics concepts. This course serves students from all NSM majors, allowing students who may not have a major or minor in physics to gain a strong content knowledge for teaching physics, especially in high-needs schools which may have them assigned to teach courses outside of their expertise.

As a result of our collaboration, the course was institutionalized as an integral part of the *teach*HOUSTON program and was the second to be approved by the Physics Department Undergraduate Studies Committee for use as an advanced physics elective in the BS/BA degree programs with the *teach*HOUSTON option. Through the physics and teacher education collaboration, we hope that preservice teachers will develop deeper conceptual understanding of targeted physics concepts, create a coherent conceptual model of these concepts, and develop an understanding of how the process of guided inquiry interacts with student learning.

NOYCE CONFERENCE

Through our collaboration, we were afforded the opportunity to attend and present at the regional Noyce Conferences as well as other STEM/Education conferences. By copresenting, we shared our research findings/innovations in teaching and offered our expertise in physics content and teaching to the audience. On average, 10 Noyce-supported students from *teach*HOUSTON have attended the annual regional conference in the past 6 years and have presented alongside us as well as individually. Attending the regional Noyce conferences exposes our students to professionals in the field, thus building an extended support network and deepening their knowledge, skill, practices, and dispositions for teaching STEM subjects. Oral and/or poster presentation helps them to gain confidence and motivation for instantiating new techniques in their current and/or future classrooms. Also we were allowed to present specific details about our program's success, including the newly instituted Physics by Inquiry course, which could be replicated by other institutions. Presentations covered topics including the conversion of "cook book style" lessons to inquiry style teaching, animating lesson plans in STEM classrooms with real-world students, and growth in Mindset strategies for STEM education classrooms (Evans, McAllister, & Stokes, 2017; McAllister, Evans, & Stokes, 2017; Stokes, Evans, & McAllister, 2017). The experience gained by the students presenting alongside us and attending the conference sessions led them to write a group thank you note to the conference organizers. One student indicated in the thank you note that

> This conference was a learning opportunity that went above and beyond what I expected. I went in as a Noyce Scholar, but I came out as part of a community. I learned how to be a better teacher, peer, and colleague. I met some amazing individuals and witnessed wonderful presentations that introduced new aspects of being a teacher. I am so grateful to have been able to absorb the knowledge of like-minded people while also being able to share what we have discovered in our experiences in the classroom. Thank you for everything you did to make this wonderful opportunity happen!

Another student stated,

> Thank you so much for putting together such an amazing, informative conference last weekend. I cannot wait to start implementing all that I've learned into my own classroom one day. This conference has given me so many new strategies and ideas on how I can be the best STEM educator I can possibly be. Furthermore, due to the various great keynote speakers, I am ready to commit to the challenge of transforming STEM education and really emphasizing diversity as much as I can. Additionally, presenting at the conference will be an experience I will never forget. Thank you so much for the unforgettable memories and the priceless knowledge gained from this conference. I hope to come back again soon.

The knowledge gleaned by our students through collaboration on presentations with us (Stokes, Evans), attendance to sessions/workshops, and networking with other pre- and in-service teachers at the Noyce Conferences is priceless.

CERTIFIED PHYSICS TEACHER PRODUCTION

One of our main goals of the *teach*HOUSTON/physics collaboration is to increase the number of physics majors and minors graduating with a physics teaching certification. As mentioned, prior to this collaboration, our campus – a major research-intensive university – had produced zero certified physics teacher in the previous decade. Since our collaboration, on average the Department of Physics has graduated three physics majors/minors who are certified to teach physics per year since 2014 as outlined in Table 1. This is largely due to the relationship that was struck between physics and *teach*HOUSTON faculty and the support of the NSF Noyce Scholarship program grant. Academic and cocurricular activities of the program led to the development of major/teacher

Table 1. Graduation Numbers for Physics BS and Teacher Education Students for 2014–2019 by Academic Year.

Graduates	2014–2015	2015–2016	2016–2017	2017–2018	2018–19
Physics bachelor's degrees awarded	9	12	17	25	25
Physics teacher education graduates[a]	3	2	2	4	2

[a]Any student who graduated from with a major or minor in physics and completed the *teach*HOUSTON teacher certification program.

certification degree plans and teacher education physics content courses, which have become integral parts of the teacher certification and physics major programs, for sustainability.

PHYSICS AND TEACHER EDUCATION COURSE DEVELOPMENT

Our collaborative efforts have influenced additional programming in the Department of Physics and *teach*HOUSTON programs. The Physics by Inquiry course structure was used to develop and implement an inquiry style section of the introductory General and University physics courses and to create lessons used for the Department of Physics' Science Teacher Equity Project (STEP), which provides professional development that supports inquiry-based active learning to in-service teachers who may/may not be qualified (hold a BS or minor degree in physics) to teach high school physics. In addition, a similar course for preservice middle school teachers, "Physics for Pre-Service Teachers," has been developed in *teach*HOUSTON and has been brought to fruition in the College of Education. This course aims to build content knowledge and provide promising practices in pedagogical approaches for teaching physics.

PHYSICS TEACHER EDUCATION COALITION FELLOWS

This collaboration between us helped to build a stronger relationship with physics faculty for improving the physics teacher education program. In Fall 2017, we, along with two additional physics faculty, applied for and were selected as members of the first group of Physics Teachers Education Coalition (PhysTEC) Fellows. The PhysTEC Fellows program selects faculty for a two-year appointment from institutions across the United States to recognize and support their efforts to develop and improve physics teacher education programs. As PhysTEC fellows, the physics and *teach*HOUSTON faculty work to develop and instantiate plans that will grow the collaboration between physics and *teach*HOUSTON and improve the physics education program. The exposure received during the two-year fellow period connected our team with physics education preparation leaders and hence promoted the physics/*teach*HOUSTON program visibility within the physics teacher education community and provided us with information about tools and strategies used by other institutions for increasing the number of qualified physics teachers. As fellows, through our *teach*HOUSTON/physics collaboration, we were able to determine strengths and weaknesses of the physics teacher preparation program employing the Physics Teacher Education Program Analysis (PTEPA) Rubric (Chasteen, Scherr, & Plisch, 2018), which is a new instrument developed through PhysTEC to characterize the structure and practices of physics teacher education programs. The rubric allowed us to take an in-depth look at the current programming and the involvement of faculty from both physics and *teach*HOUSTON and how this impacts growth and improvement in the production of certified physics teachers. Results from the rubric indicated that the physics/*teach*HOUSTON program was exemplary in program team attributes which attests to the strong relationship between us and the faculty from our respective departments. The rubric also showed that *teach*HOUSTON/physics was exemplary for its early exposure to teaching, practical K-12 experience, streamlined program accessible options, and mentoring, community, and professional support. An area for improvement suggested more physics faculty involvement in recruiting efforts and in-service mentoring. We continue to utilize the PTEPA Rubric as a tool for identifying areas that are strengths and to pinpoint areas needing programmatic improvement.

AWARDS

The outcomes of our collaborative efforts have also been realized through several awards. The Department of Physics applied for and was awarded the American Physical Society Committee on Education 2019 Improving Undergraduate Education Award for the Department's focus on retention of underrepresented minorities and improvements in the physics teacher education program. Also, we were a part of group of researchers, including three faculty from the College of Natural Sciences and Mathematics and the Colleges of Education, who received the 2019 American Educational Research Association (AERA) Narrative Special Interest Group Outstanding Publication Award for the article "*The embodied nature of narrative knowledge: A cross-study analysis of embodied knowledge in teaching, learning, and life*" (Craig et al., 2018). The lead author for the paper, Dr. Cheryl Craig, Professor and the Houston Endowment Endowed Chair in Urban Education at Texas A&M University, served as the program evaluator for the Noyce and other scholarship programs where she recognized the common theme in the interview and survey data collected over the period of the Noyce Scholarship Program related to how scholars' life experiences impacted how they learned. The article analyzes the prior knowledge and experiences of the students through their stories of their educational journeys. This collaborative work is very important as Stokes affirms that "…embodied knowledge influences how students see themselves as a scientist, a teacher of STEM, or as someone who is moving forward in a STEM career."

SUMMARY

Through the efforts of the physics and the *teach*HOUSTON faculty, a collaborative model has been developed which can be used nationwide for implementing and/or improving physics teacher education programs. This chapter highlights the importance of collaboration, trust among departments to know that courses content taught in education and vice versa are of top quality to satisfy physics and teacher certification requirements, and the impact of the collaboration of certified physics teacher production. The program, which was a result of innovative ideas put into concrete action, prepares physics teachers utilizing inquiry-based teaching pedagogies in the Physics by Inquiry course and increases the effectiveness of teaching methods so that the teachers can teach their students using methods that are most productive. The NSF Noyce Scholarship program allowed the team to build on the success of the existing teacher certification program and to develop and introduce new courses and cocurricular activities to build preservice teachers' content knowledge, professional skills, and self-efficacy, all of which are needed to retain physics teachers to the teaching profession. We will continue to collaborate and explore ways to improve, grow, and sustain the *teach*HOUSTON/physics teacher education program and invite more physics faculty to become involved in physics teacher education to prepare preservice teachers who will teach students who will attend our university (UH) in the future.

AUTHORS' NOTE

This research was funded by the following grants:

- National Science Foundation DUE Award 1240083, *Recruitment, Preparation and Retention of STEM Students as High School Teachers*
- National Science Foundation DUE Award 1557309, *University of Houston: Learning through Informal and Formal Experiences*
- National Science Foundation DUE Award 1557273, *Collaborative Research: Understanding Robert Noyce Teacher Scholarship Outcomes in Texas*

REFERENCES

Arons, A. B. (1976). Cultivating the capacity for formal reasoning: Objectives and procedures in an introductory physical science course. *American Journal of Physics*, *44*, 834–848.

Chasteen, S. V., Scherr, R. E., & Plisch, M. (2018). *A study of thriving physics teacher education programs: Development of the physics teacher education program Analysis (PTEPA) Rubric*. College Park, MD: American Physical Society.

Craig, C., You, J., Zou, Y., Verma, R., Stokes, D., Evans, P., & Curtis, G. (2018). The embodied nature of narrative knowledge: A cross-study analysis of embodied knowledge in teaching, learning, and life. *Teaching and Teacher Education*, *71*, 329–340.

Cummings, K. (2011). A developmental history of physics education research. National Academy Press.

Evans, P., McAllister, L., & Stokes, D. (2017). Implementing the 5E lesson plan in STEM classrooms. Western Regional Noyce Conference, Fresno, CA. (with Noyce scholars).

Fuller, R. (2003). "Don't Tell Me, I'll Find Out" Robert Karplus: A science education pioneer. *Journal of Science Education and Technology*, *12*(4), 359–369.

Institutional Research, University of Houston. (2019). Fall 2019 facts. Retrieved from https://www.uh.edu/ir/reports/facts-at-a-glance/facts-at-a-glance.pdf

McAllister, L., Evans, P., & Stokes, D. (2017). State of mind: Integration of growth mindset strategies in STEM education classrooms. Western Regional Noyce Conference, Fresno, CA. (with Noyce scholars).

McDermott, L. C. (1991). Millikan award lecture: What we teach and what is learned—Closing the gap. *American Journal of Physics*, *59*, 301–315.

McDermott, L. C., Heron, P. R., Shaffer, P. S., & Stetzer, M. R. (2006). Improving the preparation of K-12 teachers through physics education research. *American Journal of Physics*, *74*, 763–767.

National Academy of Sciences, National Academy of Engineering, and Institute of Medicine. (2007). *Rising above the gathering storm: Energizing and employing America for a brighter economic future*. Washington, DC: The National Academies Press. doi:10.17226/11463

National Science Board. (2000). A history in Highlights 1950-2000. Retrieved from https://www.nsf.gov/nsb/documents/2000/nsb00215/nsb50/1970/pol_dir.html

National Science Foundation. (2020). About the national science foundation. Retrieved from https://www.nsf.gov/about/

Stokes, D., Evans, P., & McAllister, L. (2017). *From cookbook to inquiry in STEM classrooms*. Western Regional Noyce Conference, Fresno, CA. (with Noyce scholars).

Tesfaye, C. L., & White, S. (2012). *High school physics teacher preparation*. College Park, MD: American Institute of Physics. Retrieved from http://www.aip.org/statistics/reports/high-schoolphysics-teacher-preparation. Accessed on January 21, 2021.

Texas Education Agency. (2017). Texas essential knowledge and skills (TEKS). Chapter 112. Science. Retrieved from https://tea.texas.gov/academics/curriculum-standards/teks/texas-essential-knowledge-and-skills

UTeach. (2020). University of Texas at Austin Program. Retrieved from https://uteach.utexas.edu/

U.S. News & World Report. (2019). The most ethnically diverse national universities in America. Retrieved from https://www.usnews.com/best-colleges/rankings/national-universities/campus-ethnic-diversity

A NARRATIVE INQUIRY INTO TEACHING PHYSICS AS INQUIRY: ONE TEACHER'S JOURNEY

Paige K. Evans, Donna W. Stokes and Cheryl J. Craig

ABSTRACT

In order to teach science effectively, teachers need a strong background in science content as well as an understanding of productive methods of teaching. This includes inquiry-based learning that will cultivate conceptual development of science concepts with their students. Furthermore, it is imperative to use student-focused activities in high-needs schools to engage all students, particularly students of color, in the learning process. As a result, faculty from the teachHOUSTON Program and the Department of Physics at the University of Houston produced a Physics by Inquiry course to engage middle school and high school preservice teachers in interactive, inquiry-based teaching pedagogies for physics. This chapter provides an overview of the course. It also highlights the benefits of including such a course in a STEM teacher education program.

Keywords: Inquiry-based learning; preservice teacher education; physics teacher education; STEM teacher education program; inquiry-based teaching; *teach*HOUSTON

Education is the tool that enables students to get ahead in life. However, will students be able to get a step ahead if they begin a step behind? Concerning the purposes of schools, Ravitch (2010) informs us that there are "primary mechanism [s] through which a democratic society gives its citizens the opportunity to attain literacy and social mobility" (p. 6). Nevertheless, Darling-Hammond (2010) maintains, "Today, in the United States of America, only 1 in 10 low-income

Preparing Teachers to Teach the STEM Disciplines in America's Urban Schools
Advances in Research on Teaching, Volume 35, 41–64

ISSN: 1479-3687/doi:10.1108/S1479-368720210000035004

kindergartners becomes a college graduate" (p. 3). In fact, students in public schools receive vastly different educations. Education reform efforts in the United States, specifically, in the STEM fields, are integral for the United States to remain globally competitive (National Academy of Sciences, 2007; National Research Council, 2011, 2012). Moreover, expert teachers and leaders are essential to improving STEM education, which requires substantial investments in teacher quality. Therefore, it is essential to work toward helping educators embrace student-focused methods that prepare all students to become productive members of society (Darling-Hammond, 2012). Although teaching-science-as-inquiry has been widely recommended (American Association for the Advancement of Science, 1993; National Academy of Sciences, 2007; National Research Council [NRC], 2000, 2005) and the importance of utilizing inquiry approaches when teaching science is addressed through both preservice and in-service science teacher preparation, research has shown that the majority of teachers fail to include inquiry teaching methods into their teaching repertoires (Goodrum & Rennie, 2007; MacLeod, 2013; Silm, Tiitsaar, Pedaste, Zacharia, & Papaevripidou, 2017). Moreover, this may occur even less in high-need and rural schools (Townsend, McKinnon, Fitzgerald, Morris, & Lummis, 2017). Nevertheless, studies show that teachers who have experienced inquiry are more likely to practice the inquiry method in their own classrooms (McDermott, 2007; Pereira, 2005; Windschitl, 2002).

The Physics by Inquiry course of the *teach*HOUSTON program provides a means for preservice teachers to learn physics through inquiry and to prepare to teach students in their future classroom in the same manner. This chapter illuminates the impact that an undergraduate level physics by inquiry teacher education course had on the learning and teaching styles of secondary mathematics and science preservice teachers in the Greater Houston area. The knowledge gleaned from the research on the course has led us to ascertain the influence of inquiry-based instruction for learning physics content; determine whether preservice teachers increase their understanding of inquiry-based learning; determine if these preservice teachers are more apt to employ inquiry-based learning in their classroom; and unearth roadblocks encountered when altering mandated curriculum in public schools to include inquiry-based teaching/learning. Outcomes and data from the course were highlighted and analyzed to inform educational practices in preparing undergraduate science teachers in diverse, urban settings. As coauthors, we begin with a review of literature and our research method. Stories of experiences (Connelly & Clandinin, 1990) of student teachers as they enact inquiry-based lessons in their classrooms will be featured. This will include one teacher's journey, over the course of five years, as he transitioned from student teaching to independent teaching in his own classroom. A summary is provided to discuss implications and future steps of utilizing a physics inquiry course for preservice teacher education.

LITERATURE REVIEW

As we look at the impact of the Physics by Inquiry course on the *teach*HOUSTON program, we consider literature which addresses the following three pillars of the research related to the theoretical framework, outputs, and outcomes: (1) inquiry-based learning; (2) science/physics reform; and (3) potential roadblocks.

Inquiry-based Learning

Inquiry began with Dewey (1933, 1938, 1944), who called for active learning on the part of students and likened the classroom as a metaphor for the laboratory of life, and continued with Schwab (1960a, 1962), who drew on Dewey and believed that students should employ the same processes that scientists engage in to discover scientific concepts. As early as 1916, Dewey maintained that science concepts are best taught through the inquiry method. It took 40 years for this initiative to develop into a large-scale curriculum development movement when the National Science Foundation sponsored several curriculum development projects including the Biological Sciences Curriculum Study, the Physical Science Study Committee, and the Earth Science Curriculum Project (Lawson, 2002). These projects led to the development of many professional development sessions that guided teachers on teaching with the inquiry process. However, even when pre- and in-service science teachers are prepared through professional development sessions to teach using promising practices, such as that of inquiry, many resort to teaching in ways that conflict with this preparation. In fact, many teachers resort to teaching the way they were taught (McDermott, 2007; Olson, 1995; Pereira, 2005; Windschitl, 2002).

Research shows that many teachers have preconceived notions regarding methods to present curriculum to a class based on their own experiences as a student. Windschitl (2002), states that "If teachers are willing to 're-culture' these kinds of classrooms, their first obstacle is the influence of their own personal histories as learners. Most teachers are themselves products of traditional schooling" (p. 151). Undoing several years of learning in a lecture type format is a challenge that must be undertaken if teachers are to feel comfortable with a different type of teaching style such as that associated with inquiry-based teaching/learning. "Teachers are more likely to be guided not by instructional theories but by the familiar images of what is proper and possible in the classroom settings" (Elbaz, as quoted by Windschitl, 2002, p. 151). Since many teachers are products of a lecture-type learning environment, they may not be familiar with the inquiry method except for what they have read in a book, studied briefly in college, or experienced through professional development sessions. Learning through inquiry should not be limited to student learning as teachers can also learn scientific concepts through inquiry-based learning. In order to dispel the lecture-type image that many teachers possess, teachers can learn curriculum content through the inquiry method which will serve two purposes: (1) they may have a deeper conceptual development of the concepts by their own "step-by-step" conceptual development; (2) they will experience the inquiry method firsthand, thus being able to change their mental model of what an effective classroom should look like.

Teacher as Curriculum Maker versus Teacher as Curriculum Implementer

In order for teachers to enact inquiry in their classrooms, it is important to engage in curriculum making with their students – that is, tailor the lived curriculum to meet their students' needs. Teachers in the United States start with standards from which to teach. In Texas, teachers follow the Texas Essential Knowledge and Skills – TEKS (Texas Education Agency [TEA]), which are the state standards that guide the concepts students should understand as well as the process skills needed to obtain scientific knowledge. Moreover, many teachers are given a binder with a complete curriculum document to implement in their own classroom. Those teachers that take into account their students and their environment while deciding how to teach a particular concept act as curriculum makers.

> The teacher as curriculum maker image works from the assumption that a classroom space exists within which teachers and students negotiate curriculum unhampered by, though not oblivious to, others' mandates and desires. (Craig, 2010b, p. 868)
>
> It is a view in which the teacher is seen as an integral part of the curricular process and in which teacher, learners, subject matter, and milieu are in dynamic interaction. (Clandinin & Connelly, 1992, p. 392)

In Craig's view (2010b), the

> ...teacher as a curriculum maker is an image that acknowledges the teacher as a holder, user, and producer of knowledge, a self-directed individual who takes the curriculum as given and negotiates it in active relationship with students to address their needs as learners and, to the extent possible, meet the requirements outlined in stated curriculum documents. (p. 867)

Those teachers that merely execute the curriculum as stated are known as curriculum implementers and are described as follows:

> In the [teacher-as-curriculum-implementer] conceptualization, the teacher uses other people's knowledge and, in a technical rational way, installs a curriculum/curriculum package designed by others. In short, the image of teachers as curriculum implementers treats teachers as functionaries who are totally reliant on state and national imperatives. (Craig, 2010b, p. 869)

For curriculum making to occur, Schwab (1960b/1978) maintained that there are four bodies of experience that must be represented as shown in Fig. 1. These four commonplaces include the following: subject matter, learners, milieus, and teachers. As Craig explains (2011):

> Curriculum is what happens – what becomes instantiated – in the moments when teaching and learning fuse. In that fusion, teachers use what is in their students (learner commonplace), their teaching situations (milieu commonplace) and themselves (teacher commonplace) to make curriculum (typically organized around the subject matter commonplace) in a way that cannot be captured in a codified knowledge base without negating the continuity of experience (Dewey, 1938) fueling the human knowing. (p. 2)

Thus, inquiry naturally causes the teacher and the students to engage in curriculum making which allows for a deeper understanding of physics concepts. It also minimizes methods of teaching not conducive to deep conceptual development such as memorization, rote learning, and using textbooks as the

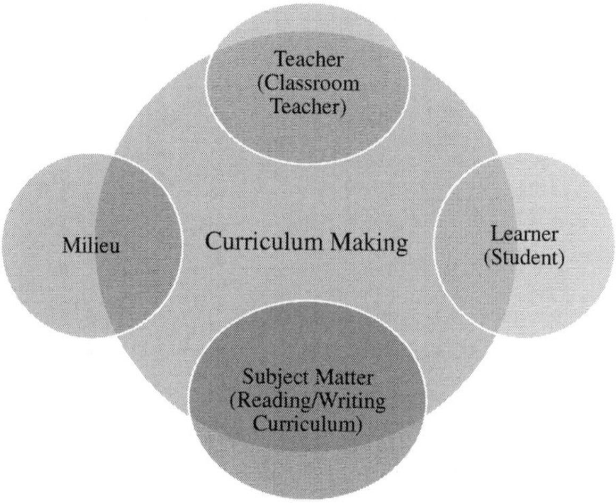

Fig. 1. Schwab's Curriculum Commonplaces. *Note:* Milieu includes local school district policies, state educational policies, national educational policies, etc.

curriculum. Furthermore, it may put the learning at the forefront of education instead of basing the curriculum on standardized testing.

Reforms in Teaching Science/Physics

As mentioned earlier, teaching science as inquiry has shown to be effective and leads to improved learning on the part of the students (Marshall, Smart, & Alston, 2016; Meltzer & Thornton, 2012). The National Science Education Standards (National Research Council, 1996) defines inquiry as the following:

> Scientific inquiry refers to the diverse ways in which scientists study the natural world and propose explanations based on the evidence derived from their work. Inquiry also refers to the activities of students in which they develop knowledge and understanding of scientific ideas, as well as an understanding of how scientists study the natural world. (p. 23)

The National Research Council (2000) depicts the Essential Features of Classroom Inquiry and Their Variations in Table 1.

When students inquire, they are constructing their own knowledge and are active in the learning process. Active engagement in the learning process has always been crucial to concept development. Fundamental inquiry-based learning is the active involvement of students in the learning process and that students are "doing rather than being told or only reading about science" (NRC, 2000, p. 16). Thus, teachers are curriculum making along with their students. Physics education research also deems inquiry-based learning pivotal to student success. Much research (McDermott, 1993, 1996, 2007; McDermott, Shaffer, & Constantinou, 2000; McDermott & Shaffer, 1992) has been done on how students best learn physics concepts. According to Dubson (2007), "Trying to learn

Table 1. Essential Features of Classroom Inquiry and Their Variations.

Essential Feature	Variations			
1. Learner engages in scientifically oriented questions	Learner poses a question	Learner selects among questions, poses new questions	Learner sharpens or clarifies question provided by teacher, materials, or other resources	Learner engages in questions provided by teacher, materials, or other sources
2. Learner gives priority to **evidence** in responding to questions	Learner determines what constitutes evidence and collects it	Learner directed to collect certain data	Learner given data and asked to analyze	Learner given data and told how to analyze
3. Learner formulate **explanations** from evidence	Learner formulates explanation after summarizing evidence	Learner guided in process of formulating explanations from evidence	Learner given possible ways to use evidence to formulate explanation	Learner provided with evidence and how to use evidence to formulate explanation
4. Learner connects explanations to scientific knowledge	Learner examines other resources and forms the links to explanations	Learner directed toward areas and sources of scientific knowledge	Learner given possible connections	
5. Learner communicates and justifies explanations	Learner forms reasonable and logical argument to communicate explanations	Learner coached in development of communication	Learner provided broad guidelines to use to sharpen communication	Learner given steps and procedures for communication

More ------------ Amount of Learner Self-Direction ------------ Less
Less ------------ Amount of Direction from the Teacher or Material ------------ More

Source: (NRC, 2000, p. 23).

physics by watching your instructor do physics is like trying to learn piano by watching your piano teacher play. It is remarkably ineffective" (p. 252). Studies have shown that time utilized in the classroom for hands-on investigations coupled with leading questions and meaningful discussion can eliminate many misconceptions in physics (McDermott & Shaffer, 1992). Additionally, Redish and Steinberg (1999) concluded that students need to be more actively engaged in learning rather than just listening to a professor lecture and watching him/her solve difficult physics problems. As maintained by Jasien and Oberem (2002), in-service teachers learned more when using Physics by Inquiry curriculum (McDermott, 1996) during a three-week professional development than teachers that learned through traditional lecture. McDermott et al. (2000) maintain that teachers need to have a deep conceptual understanding of physics concepts. As a result, they need to be afforded the opportunity "to understand not only what we know, but on what evidence and through what lines of reasoning we have come through this knowledge" (p. 412). As maintained by McDermott et al. (2000),

> Science instruction for young students is known to be more effective when concrete experience established the basis for construction of scientific concepts. We and others have found that the same is true for adults, especially when they encounter a new topic or a different treatment of a familiar topic. Therefore, instruction for prospective and practicing teachers should be laboratory-based. (p. 411)

Such thinking resonates with Dewey's (1938) theory of experience and the importance of the laboratory of life and with Schwab's notions of inquiry as a "generic envelope" (Schwab, 1962), both of which pattern how scientists think. Nevertheless, novice and veteran teachers often have difficulty enacting inquiry-based learning into their physics and science classrooms because of roadblocks including their own mindset, their lack of preparation, and the structure of the educational system which are highlighted in the next section.

Potential Roadblocks

Although utilizing the scientific inquiry as a way of obtaining scientific content knowledge seems feasible, there are several roadblocks to achieving this goal. One barrier to enacting inquiry lessons is that teachers may have a difficult time letting go of the role of being the disseminator of knowledge since they may have been inducted into that mode by their education, i.e., many teachers have preconceived notions regarding a proper class based on their own experience as a student. "Teachers tend to teach as they were taught. If they were taught through lecture, they are likely to lecture, even if such instruction is inappropriate for their students" (McDermott et al., 2000, p. 412). In addition, teachers may be expected to play the role of the disseminator by those directing educational policy at school district, state, and national levels.

Another potential roadblock to teaching through inquiry is the lack of preparation of the teacher. Effective teaching requires a strong background in science content as well as implementation of promising practices in teaching that will cultivate conceptual development of science concepts for the students. Research has been performed to determine how students best learn science concepts and active engagement in the learning process has always been identified to be crucial to concept development as emphasized by Dewey, Schwab, and Piaget. Inquiry lessons, coupled with constructive discussion among students in collaboration with the teacher, help students cultivate connections between ideas; however, facilitating inquiry-based lessons requires the teacher's in-depth knowledge of the subject at a time when many teachers are teaching out of their content areas. For example, approximately 30% of high school science teachers in Texas are teaching out of their fields (Texas Education Agency, 2017). And, approximately 30% of teachers in Texas are certified by Alternative Certification Programs (Ramsay, 2019), most of which do not have field experiences as part of their program.

The structure of the educational system may be one of the most critical roadblocks. Dewey (1938) argued that the school system is set up for a more traditional form of education with the patterns of the typical school consisting of exams, schedules, and bells within the school. Dewey (1938) stated the following about traditional education,

> The subject matter of education consists of bodies of information and of skills that have been worked out in the past; therefore, the chief business of the school is to transmit them to the generation. (p. 17)

Perhaps Dewey's pedagogy would be more prevalent in today's curriculum if the drive to create standardized knowledge statements (standards) and standardized testing, both of which rely on formal knowledge claims, had not put a halt to his ideas which also took into account personal expressions of knowing. As Kliebard (2004) explains:

> The rise of standardized achievement tests in the twentieth century would sharply accelerate the tendencies in the teaching of the three R's that he so much deplored and would help make his own emphasis on the relationship between reading and human purposes of the object of scorn and caricature. (p. 68)

Since most high stakes tests are made from educational standards, the vicious cycle of standardized tests is reducing the quality and quantity of education offered to the children of Texas with the most damaging effect on minorities and low socioeconomic youth (McNeil & Valenzuela, 2001). In short, testing rather than physics, biology, mathematics, or chemistry has supplanted curriculum and practical and formal knowledge gleaned from curriculum. Studies have shown the high stakes testing in general have had a detrimental effect on our education system. As indicated by Nichols and Berliner (2005), implementing high stakes testing has negatively impacted every level of the public school system. Moreover, teachers are not empowered to play an integral part as curriculum makers informing the curriculum. Darling-Hammond (1997) asserts that "A coherent view of curriculum, assessment, and teaching is at the core of any vision of more effective education" (p. 211). While this may be true, in the US educational system the profession has turned into a huge bureaucracy where teachers have little or no say on "standard setting, curriculum development, or assessment" (p. 213). This vicious cycle of curriculum design based on what is easily testable and what power brokers rather than what society wants and needs is well documented throughout history. Dewey also felt the disconnection associated with developing standards, curriculum, and assessment a century ago. He was of the belief that teachers who have "large experience and riper wisdom" should be tasked with how "life comes to [youth]" (Dewey, 1897, p. 79). Dewey (1897) nailed the underlying issue with policy-related reforms in his pedagogical creed in the 1800s, "…reforms which rest simply upon… enactment of law, or… threatening of certain penalties, or upon changes in mechanical or outward arrangements, are transitory and futile" (p. 16).

> Dewey also called attention to "the mechanics of school organization and administration" that they often control what gets taught. (p. 74)

Should money and resources continue to be invested on professional development, particularly for inquiry teaching/learning, when teachers are unable to apply what they learned in the classroom or have any say in curriculum writing? The classroom is where the research, collaboration, and curriculum planning should take place. According to Clandinin and Connelly (1992), the teacher, researcher, and curriculum specialist must work together and all be equal in the process in order for curriculum writing to be successful. Clandinin and Connelly note,

> The future direction for the teacher as curriculum maker is an agenda shared with teachers. It has two parts: to listen to teachers' stories of their work and to construct together stories of teachers as curriculum makers. Both parts send us back to the classroom to listen to teachers' stories of their work. (p. 392)

Effective research can impact our educational system and inform practice if the teacher is an integral part of the process; however, as Dewey purported, "as long as the system of rewards remains the same, the reform is doomed" (Kliebard, 2004, p. 74). In other words, as long as the emphasis remains on standardized testing, the teacher may not have much say in the curriculum. Schwab (1983) viewed teachers as "the fountainhead of the curriculum decision" because only he/she meets the students face-to-face. To him, "there are a thousand ingenious ways in which commands on what and how to teach can, will, and just be modified or circumvented in the actual moments of teaching" (p. 245). In Schwab's words,

> Teachers practice an art. Moments of choice of what to do, how to do it, with whom and at what pace, arise hundreds of times a school day, and arise differently every day with every group of students. No command or instruction can be so formulated as to control that kind of artistic judgment and behavior, with its demand for frequent, instant choices of ways to meet an every varying situation. (p. 245)

Schwab believed it was necessary for teachers to be involved in decisions regarding what must be taught and how it should be taught. Although today's educational system is fraught with the aforementioned roadblocks, some teachers are able to overcome these challenges and infuse inquiry-based learning into their teaching repertoires. Behind the closed doors of the classrooms where they have discretionary space (Craig, 2009), they can seize the possibilities of their situations, grab the opportunity to teach using inquiry methods, and "bring life to [youth]" in ways that will help the young people before them to replenish and sustain their learning selves in the future.

METHODOLOGY

Narrative Inquiry

Using the method of narrative inquiry (Clandinin & Connelly, 1992, 2000; Craig, 2011; Polkinghorne, 1995), an inquiry into the narratives of one physics teacher who began as a preservice teacher who was taught physics through inquiry and who now teaches physics in a public school in Houston, the fourth largest urban center in America is highlighted. Narrative inquiry was strongly influenced by John Dewey (1938) who believed that one must rely on past experiences and knowledge to solve problems and that life experience is in fact education. Furthermore, Clandinin and

Connelly (2000) purport that "Experience happens narratively. Narrative inquiry is a form of narrative experience. Therefore, educational experience should be studied narratively" (p. 19). The process of narrative inquiry involves "a collaboration between the researcher and the participants, over time, in a place or series of places, and in social interactions and with milieus" (p. 20). Narrative inquiry occurs within a *three-dimensional narrative inquiry space* which allows us to travel in the following directions: "inward, outward, backward, forward, and situated within the place" (Clandinin & Connelly, 2000, p. 49).

This narrative inquiry studied a preservice teacher as he learned physics content through inquiry and how to craft inquiry-based lessons as a student in the *teach*HOUSTON teacher education program during student teaching and later as a physics teacher. His story expresses the experiences he encountered in the *teach*HOUSTON physics inquiry course and subsequent enactment of inquiry-based learning during his first year as a teacher. The future was discussed as a reflection on how to incorporate what was learned to strengthen the practice of preparing preservice mathematics and science teachers to teach in diverse, urban settings. His narratives form the primary data source for exploring the impact of learning through inquiry and the impact it has on the teacher's teaching style used in the classroom. To begin, a description of the infrastructure of the *teach*HOUSTON teacher preparation program, which offers the physics inquiry course, is presented along with a description of the NSF grant, and then the physics inquiry course.

Context of the Inquiry

teachHOUSTON Program

UH's existing *teach*HOUSTON program (*teach*HOUSTON Program, 2017) is the first replication site of UTeach, a nationally acclaimed teacher preparation program (Uteach website, 2017). *teach*HOUSTON is a collaboration between the College of Natural Science and Mathematics (NSM), COE, and local school districts and aims to combat the shortage of qualified math and science teachers (National Academy of Science, 2007). *teach*HOUSTON, which started in 2007 with 14 students, currently enrolls approximately 370 students. The program aligns with the goal of the *Prepare and Inspire* report to prepare 100,000 new quality STEM middle school and high school teachers (President's Council of Advisors on Science and Technology, 2010). The program provides grades 7-12 teacher certification for students obtaining an NSM degree while emphasizing early and ongoing field experiences. To date, *teach*HOUSTON has graduated 347 secondary STEM teachers that are changing the region's public education systems: approximately 90% go directly into teaching; 88% of the graduates are teaching in the Greater Houston area; 80% of the graduates teach in high-need schools; 87% continue beyond five years of teaching; and the program produces one of the country's largest pools of chemistry teachers (top 2%). Moreover, the *teach*HOUSTON graduates mirror local diversity, which is a stark contrast to the Texas and US teacher populations (NCES Indicator, 2018; TEA, 2018). This composition is fundamentally important because successful academic outcomes more likely occur when minority teachers serve as role models (Basit & McNamara, 2004; Dee, 2005; Pitts, 2007).

NSF Noyce Grant

A joint effort has been made by *teach*HOUSTON and the Departments of Physics and Education to combat the shortage of qualified physics teachers. Through a National Science Foundation Noyce Scholarship program with curricular activities and the development of major/teacher certification degree plans, the program has graduated 13 certified physics teachers as of 2019 where prior to this effort, UH had not graduated any students certified to teach physics in over a decade.

The scholarship program, specific to physics and chemistry majors/minors, focuses on increasing the knowledge base around two key elements of STEM preparation: (1) Developing effective methods to attract and retain STEM majors into secondary education and (2) incorporating key principles of research in education and learning sciences into STEM coursework to transform undergraduate teaching and learning. The program provides a scholarship of $12,000 for up to two years in return for a teaching commitment in a high-need school district for 2 years for each year of scholarship. The program also includes professional development activities, mentoring, cohort building through *teach*HOUSTON, and the aforementioned "Physics By Inquiry" course all of which have become integral parts of the teacher certification and major programs, making them sustainable even if the scholarship support is not available.

Physics Minor

Although the Noyce Scholarship program has been instrumental in attracting physics majors to *teach*HOUSTON, retaining them has been accomplished through the establishment of degree plans for a BA/BS degree in physics with a *teach*HOUSTON minor and a BS degree in Mathematics with a physics minor. These plans are designed for degree completion and teaching certification within 4 years, making them attractive to STEM majors interested in teaching. These specialized degree programs were approved and promoted by both the Physics Department and *teach*HOUSTON. Additional details on these degree programs can be found in Chapter 3 Collaboration between a Physics Professor and a Physics Teacher/Teacher Educator.

The Physics Inquiry Course

Research (Lederman, Gess-Newsone, & Latz, 1994; McDermott, 1984; Putnam & Borko, 1997) shows that content knowledge of a subject is critical for effective teaching of science content; therefore, physics teachers will need to have a deep conceptual understanding of the physics principles as well as possess quantitative skills for solving problems and analyzing data needed to teach their students. Physics teacher preparation begins in their physics major degree program and continues through their teacher education training (Feiman-Nemser, 1983). Preservice teachers complete courses required for a physics degree which mostly aims to train students to pursue a future in a technical field. Frequently, these programs include courses which may not necessarily be an asset to a student who is preparing to teach physics at the secondary level. For many preservice teachers, the physics content is mostly taught to them by faculty in physics departments for their undergraduate degree programs which is oftentimes taught in a traditional lecture style versus an interactive inquiry style. This results in preservice teachers experiencing physics learning associated with mastering the fact/concepts, applying them to physics problems through problem-solving and lab experimentation, and then being tested on their mastery (Anderson & Mitchener, 1994). Researchers questioned if this type of instruction was suitable for effectively preparing physics teachers (McDermott, 1990). In McDermott's analysis of the instruction preservice teachers receive in their University Physics Courses (calculus-based physics course sequence completed by all physics majors), she determined that these courses do not provide the necessary preparation required for teachers, particularly for teaching by inquiry (McDermott et al., 2000). Her research promoted the inclusion of a physics course geared toward preparing teachers which utilized inquiry style teaching/learning taught within the Physics Department. Despite this recommendation, this practice is not the choice of most physics departments. Historically, many physicists feel that teacher education courses should be offered and taught by faculty in teacher education programs which are traditionally housed in Colleges of Education. However, with the creation of the UTeach Program, at the University of Texas in Austin, STEM teacher education programs are now finding a home in Colleges of Natural Science making it easier to provide preservice teachers with content-specific courses which offer the science content using promising practices for delivery. This also allows for incorporation of curriculum and pedagogy development to be included within the course content. Despite having UTeach and its replication sites at 46 institutions (this includes *teach*HOUSTON) across the United States, many physics as well as teacher education faculty do not agree that the courses, i.e., science-based teacher education or science courses, are adequately preparing qualified physics teachers but do agree that preservice teachers need to be well trained for teaching high school students who may end up in their courses (Azam, 2018; Razali & Yager, 1994; Sadler & Tai, 2001).

The *teach*HOUSTON program did not have a physics-specific course for preservice teachers; however, the Physics Department did have, but no longer offers, several courses for in-service physics teacher content training. In the early years of *teach*HOUSTON, there was no collaboration between the *teach*HOUSTON and the Physics Department faculty for ensuring effective proper training of physics teachers. Realizing the importance of collaboration and a physics teacher preparation course, Physics faculty (Stokes) and *teach*HOUSTON faculty (Evans) worked to develop and institutionalized a physics teacher education course as part of the Physics and *teach*HOUSTON course listings.

The Physics by Inquiry course was developed based on promising practices of the use of inquiry for teaching/learning physics concepts. The curriculum is a laboratory-based course in which preservice teachers develop a deep understanding of physics concepts by experimenting and establishing scientific concepts based on a concrete experience. This physics course utilizes research-based, effective pedagogical approaches to teaching science including guided inquiry, student-centered learning, group discussions, demonstrations, lectures, lab investigations, and independent research. The goal is for students to develop a deep conceptual and mathematical understanding of physics content. Through inquiry-based learning, students develop scientific critical thinking and reasoning skills.

This curriculum was developed by The Physics Education Group at the University of Washington (McDermott, 1996) which has over 25 years of experience in preparing teachers to teach physics and physical science by inquiry. Evans (Chapter Author 1) and Stokes (Chapter Author 2) added their own inquiry lessons to this course that they created over the years. Topics in the course include the following:

- Electric Charge, Forces and Fields
- Electric Potential and Potential Energy
- Electric Current and DC Circuits
- Magnetism and Faraday's Law
- Waves and Sound
- Light and Optics
- Quantum and Atomic Physics
- Thermodynamics

According to McDermott et al. (2000),

> Whether intended or not, teaching methods are learned by example. If the ability to teach by inquiry is a goal of instruction, teachers need to work through a substantial amount of content in a way that reflects this spirit. (p. 413)

Additionally, the notion of separating science instruction from instruction in pedagogy diminishes the value of both for teachers. Therefore, teachers may not be able to adapt a teaching strategy to a novel situation. Consequently, learning science content as inquiry will benefit preservice science teachers as this may influence their perception of what a typical classroom should look like and increase their confidence in teaching science as inquiry.

An undergraduate physics inquiry course has been taught in the *teach*HOUSTON program since 2013 using McDermott's Physics by Inquiry Volumes I and II (McDermott, 1996) as a resource as well as other materials developed by the authors. The course was designed to augment the content knowledge of the participant preservice teachers through instructional strategies grounded in promising practices and inquiry-based teaching pedagogies, while allowing the preservice teachers to experience the process of inquiry learning. Preservice teachers' experiences in the course and its impact on their instructional practices as in-service teachers were analyzed through narrative inquiry methods as a focus of this research.

Narrative Inquiry

In this narrative inquiry, a variety of field texts were utilized (Clandinin & Connelly, 2000) including autobiographical writing, conversations, journal writing, field notes, correspondence, direct observations, documents, and life experience. Field notes were recorded daily and during observations of pre- and in-service teachers teaching their inquiry-based lessons.

Several documents were gathered throughout the study which consisted of the following: course syllabus, pre-/post-tests, lesson plans, pre-/post-surveys, and assignments. These artifacts were used to analyze improvement in content as well as to understand the dispositions participants have regarding inquiry-based learning/teaching.

Interpretive Tools

Three modes of interpretive tools were utilized to support the narrative inquiry including broadening, burrowing, and storying and restorying (Connelly & Clandinin, 1990). Broadening enabled us to expand beyond the context of the immediate study to obtain an overall picture of the data. The participant's data were organized electronically and include several artifacts which are listed in Table 1. Analyzing these data allowed for a general idea of how this participant improved in his physics content and to also ascertain general themes that emerged during his journal reflections. Burrowing allowed us to look intensively at the participant's experience and analyze his experiences for the larger meanings brought to the study. This enabled us to show how the physics content and personal practical knowledge (Clandinin, 1986) emerged from this participant and also to try and make sense of the experiences of the teacher and how this translated into his classroom. Storying and restorying enabled us to show how change happens at both the micro and macro levels. By analyzing the data, we looked for narrative threads of experience that convey the journey that this participant experienced as a result of the physics inquiry course. As Connelly and Clandinin maintain,

> As we engage in a reflective research process, our stories are often restoried and changed as we, as teachers and/or researchers "give back" to each other's way of seeing our stories. (1990, p. 9)

Data Collection: Sources and Instruments

Using these interpretive tools, the data collected will be used to address research questions related to the impact of the course on preservice teacher learning experience and on their use of inquiry in the classroom. Table 2 depicts data collection instruments/frequencies/participant numbers and Table 3 illustrates the research questions of the study and how the data address them.

Participants were asked to sign an IRB consent form and also given pseudonyms to maintain the confidentiality of the participants. Additionally, all participants were asked to member check and had access to all field texts and all research interpretations.

RESULTS AND ANALYSIS

After field texts were analyzed and converted into research texts, a comprehensive analysis of the Physics by Inquiry course was completed and then followed by a presentation of the "stories of experience" (Connelly & Clandinin, 1990). In

Table 2. Data Collection Instruments, Frequency, and Number of Participants.

	Frequency	Number of Participants
Inquiry-Based Instruction Survey	1	13
Inquiry-Based Instruction Post Survey	1	13
Teacher Interest Survey	1	13
Teacher Interest Post Survey	1	13
Journal Reflections	2	13
Journal Reflections Student Teaching	5	3
Pre-Test Circuits	1	13
Post-Test Circuits	1	13
Model for Circuits	1	13
Course Manual	1	13
Videos	19	13
Observations	5	3
Interviews	5	3
Lesson Plans	10	3

Table 3. Data Collection and Analysis by Research Question.

Research Question	How Does Learning through Inquiry Impact the Quality/Quantity of Physics That is Learned?	How Does Learning Physics through Inquiry Impact the Attitudes that Preservice Teachers Have toward Teaching through Inquiry?	Are Participants More Apt to Employ Inquiry Learning Subsequent to Learning Science as Inquiry?	What are Some of the Roadblocks Preservice Teachers Encounter in Their Experiences of Trying to Teach Science as Inquiry?
Inquiry-Based Instruction Survey		✓	✓	✓
Journal Reflections	✓	✓	✓	✓
Journal Reflections Student Teaching	✓	✓	✓	✓
Pre-Test Circuits	✓			
Post-Test Circuits	✓			
Model for Circuits	✓	✓		
Course Manual	✓			
Videos	✓	✓		
Observations		✓	✓	✓
Interviews	✓	✓	✓	✓
Lesson Plans	✓	✓	✓	✓

particular, a story that emerged from one participant, Jason, who embraced inquiry-based learning as both a learner, a student teacher, and then an in-service physics teacher, is highlighted. Stories of his learning depicted ways in which he overcame barriers in his own mindset and was able to transform what he knew about inquiry into teaching practices in face-to-face interactions with students.

Jason's Background

Jason was enrolled in the *teach*HOUSTON program as a mathematics major. Shortly after the inception of the Noyce Scholarship program, we approached Jason about pursuing a physics minor and described the new plan as well as the Noyce Scholarship Program. At this point in time, Jason was not sure if he wanted to teach or go to graduate school. However, having the Noyce scholarship opportunity coupled with his success in his field-based courses led him to continue to pursue teacher certification as well as the physics minor. Jason enrolled in the physics inquiry course which was a requirement for the physics minor.

Jason's Story

As previously mentioned, we approached Jason about adding a physics minor to his mathematics major. Jason shared that he enjoyed physics ever since taking it in high school. In his words, "I find it much more enjoyable than Math, although admittedly, it takes me a bit longer to grasp complex physics concepts than math concepts." Jason was the first *teach*HOUSTON student to embark on the specialized physics minor for *teach*HOUSTON students. As part of the physics minor, Jason was required to take the physics inquiry course. The course starts with a survey and journal reflection. In his first journal reflection, he explained the following:

> While I initially thought my knowledge in circuitry was quite strong (for the high school level), I continue to find myself meeting contradictions or holes in my understanding. For example, I tend to have a fairly easy going experience when depicting simple series and parallel circuits.

Jason continued:

> However, after an attempt of the first homework assignment, this preconception of perfect knowledge disappeared after viewing the third problem. This question simply asked to depict the physical layout of a circuit into a circuit diagram. These circuits were not "traditional" in the sense that they did not represent a simple physical layout. Alternatively, I might have been overanalyzing the problem and trying too hard to integrate my previous "unneeded knowledge."

This experiential learning was pivotal in that Jason was able to change his mental model of learning physics and also reimage his own misconceptions about effective classrooms. As previously mentioned, many teachers have preconceived notions regarding a proper class based on their own experience as a student. Even when preservice mathematics and science teachers are trained in promising practices, many resort to teaching in ways that conflict with this training; many teachers resort to teaching the way they were taught. In Pereira's (2005) view,

> All too often, my student teachers behave in ways that seem to contradict the vision they have developed, as if they had snapped back to a former state when faced with the realities of classrooms and schools. (p. 71)

Pereira believes the first step in education reform is to change the way that experienced and prospective teachers learn subject matter. Thus, Jason was actively starting to envision what inquiry instruction in the classroom would look like.

As part of the study of the physics inquiry course, preservice teachers fill out an inquiry-based pre- and post survey. Jason's response on the survey when asked whether or not inquiry-based instruction represents effective practices in secondary science instruction and if all instruction should be done in this format, he replied,

> I can agree that any introductory lesson can be inquiry-based, with the more complex portions being question-driven. However, for a new teacher, it would be rather difficult to accumulate a mass amount of such activities.

Here, he expresses concern about the amount of time needed to create inquiry lessons which he and others perceive as a common barrier to enacting inquiry. Fitzgerald, Danaia, and McKinnon (2019) report that teachers are dissatisfied with most of their professional development sessions because they are not provided with either inquiry-based curriculum to use or the time needed to create inquiry-based curriculum. When asked about the key elements of inquiry-based instruction, Jason reported the following about his idea of student engagement, "If the students of the classroom are utilizing hands-on experiments and discussing the relevant topics in their conversations." This shows a misunderstanding of what constitutes inquiry-based learning prior to taking the course. He further discussed that during inquiry-based learning,

> The teacher is helping the students by asking driving questions, not simply lecturing. The teacher would also be monitoring the classroom as the students commence to assure on task behavior.

Studies show that there is often confusion and misconceptions on the part of teachers as to what comprises inquiry-based teaching and learning (Marshall et al., 2016).

Jason reported on the post survey that he learned more on how an inquiry-based lesson could be implemented through this course than in his previous four field-based courses. Additionally, he expanded his understanding of inquiry-based learning and further purported that "this is the only course in which I experienced real inquiry-based instruction, rather than occasional 'model lessons' in other courses." As a result, Jason felt more confident in teaching physics as inquiry. He was able to transfer the inquiry method to other lessons in his field-based courses prior to student teaching. For example, he made a lesson over the photoelectric effect that he implemented in one of his field-based courses prior to student teaching.

When asked about how to strengthen the course, he said that he believes the course would be improved by devoting more time discussing how teaching strategies in this course could be implemented as teachers. For example, it would be beneficial to understand how the curriculum learning in the physics inquiry class can be translated to work in a high school classroom. Also, it would be great to discuss how teachers can teach inquiry-based lessons and still teach all of the objectives in the TEKS. Finally, he expressed that it would be wonderful to have more discussion on the power of inquiry so that preservice teachers can discuss inquiry with their future colleagues who may be resistant to this type of learning. Other preservice teachers mentioned this as well. These components were included in subsequent classes.

Teaching Experience

As mentioned earlier, Jason was the first *teach*HOUSTON candidate to embark on the specialized physics minor. When asked how the physics minor impacted his decision to pursue physics in his teaching career, he shared the following:

> My physics minor helped reinforce my desire to teach physics over math. I was able to learn about what I was required to teach at a much deeper level than what I knew while taking the classes at university. I have also found myself having to learn about other topics that I might not have taken in college such as Thermodynamics and Fluid mechanics. And I found that I truly loved learning about these topics as I taught them for the first time. And now they are among my favorite topics.

This shows Jason's appreciation of inquiring himself and the benefits of learning science as inquiry. As per Dewey (1933), learning is deep-rooted in the experiences one enters into and the knowledge that arises through a process of inquiry.

The semester after completing the Physics By Inquiry course, Jason embarked on a full-time student teaching semester. He was placed with a physics teacher in a high-need school district. The cooperating teacher favored the more traditional ways of teaching physics in the class. Since Jason was a guest in the classroom, this was one roadblock which limited his implementation of inquiry-based learning in the classroom.

Confronting/overcoming such a roadblock, i.e., by changing the mindset of the cooperating teacher in the classroom who taught in a more traditional fashion, was essential to enact inquiry-based lessons during student teaching and later on in his own classroom as a physics teacher. Additional roadblocks Jason faced included the time needed to write inquiry lessons; the amount of time needed to enact inquiry-based lessons; and staying with a team of teachers in a particular school. Moreover, the students were not used to the inquiry method of teaching and did not feel comfortable with this method of teaching. Instead, they preferred to be told what they needed to know and were concerned with what was on the exam. Despite the roadblocks, Jason persisted and was able to teach inquiry-based lessons when feasible. He was able to bring in small inquiry activities here and there where students were inspired to do the thinking. Later, when Jason was in his own classroom in a different school, he had more freedom to teach in ways that he deemed more beneficial to students such as the active learning strategies learned in the *teach*HOUSTON program and inquiry-based learning which was a focus of the physics inquiry course. When teaching in his own classroom, he was able to teach an entire unit in an inquiry-based fashion which led to more success experiences by his students. This particular inquiry unit had the same content as in the physics inquiry course where students were able to build their own mental models for circuits. He was able to use a modeling approach using the Capacitor Aided System for Teaching and Learning Electricity (CASTLE) curriculum (Wainwright, 2006) which had similar content as the physics inquiry course, but more appropriate for high school physics curriculum. Thus, he explained that he was prepared to teach this unit as he had recently been a student of inquiry with the same content. Additionally, Jason explained that it is much easier now that he is working with another teacher that is like-minded. In his current school, he explained that he teaches and plans with another teacher who has a similar teaching philosophy in that the teacher is also a curriculum maker who uses physics as inquiry learning strategies.

Jason strongly believes that his students benefit from engaging in inquiry lessons rather than learning through traditional means. He states that his favorite class to teach is AP physics 2. In his words,

> …when I first had to teach the class, I was stepping into a whole new world. I had no resources, no timeline, and very little memory on the content knowledge.

Thus, Jason was put into the shoes of his students who learn through doing, learn through inquiry. This resonates with Schwab's (1962) beliefs that science should be learned similar to the way that scientists conduct their research. Jason goes on to explain that he learned to discover what he was going to teach,

> Because I had to learn the content "on my own," I found myself enjoying it a lot more than simply trying to teach it "the way it was taught to me."

He sums it up by saying "I believe since I find myself enjoying the learning process through discovery, then students would certainly prefer that same style over traditional lecture!" Once again, this shows how Jason's mental model of effective teaching has changed from traditional lectures to learning as inquiry.

Jason currently serves as a cooperating teacher for another *teach*HOUSTON preservice physics teacher. Jason brings his expertise of inquiry-based learning with his mentee and shares his "do before you learn" philosophy. He has a very good relationship with his mentee. Typically, he teaches the first class, then his mentee teaches the second class "with his own flair." As the semester continued, Jason reported that his mentee was more comfortable in developing his own lesson. In his words,

> …as the semester went on, he became more and more comfortable with developing lessons himself, most of which I plan to use in my own lessons from here on. I believe it is important for both he and I to learn from the experience and for him mostly to see me as someone who is simply more experienced, but am in no way "superior."

Thus, his mentee was developing his own curriculum and was empowered by Jason to become a curriculum maker instead of a curriculum implementer (Craig,

2010b; Craig & Ross, 2008). His mentee was able to tailor the lived curriculum to meet the needs of the students.

Jason currently serves in a leadership position at his school as the physics team lead. He is responsible for communicating with the other physics teachers regarding numerous topics such as, scope and sequence, materials request, new lessons, as well as communicating with other physics teachers in the district. In his words, "I can't but feel privileged to have been awarded such a position!" He has also received a prestigious physics teaching award, currently sponsors a physics-related club, and created another club where students get together every other week to discuss certain interests such as Arduino programming, 3D printing, and web development. He explains that this club provides members with more "inquiry" since it allows students to work on their own interests. His future plans include mentoring additional preservice teachers and also expanding his leadership role at the district level to continue to work with other physics teachers in the district but in a larger capacity in which he can share his love of inquiry.

SCHOLARLY SIGNIFICANCE OF STUDY

The gap between knowing science as inquiry and teaching science as inquiry exists on many levels. There are preservice teachers who learn through their education courses that science concepts are best learned as inquiry but only experience a few inquiry-based lessons in their coursework. As a result, preservice teachers have a limited understanding of inquiry-based learning and have difficulty enacting inquiry-based lessons in their student teaching semester and subsequent teaching careers. It is crucial that preservice teachers are afforded the opportunity to learn physics content as inquiry for several reasons. First, they will learn physics concepts deeply and realize that inquiry-based instruction is effective after experiencing success in a course that is taught in this manner. Second, they will be able to change their mental model of how a physics classroom should operate and have more confidence employing inquiry-based lessons in their classrooms. It is important to provide preservice and in-service teachers with support, effective curriculum that they can modify, a collaborative group of peers, and receptive cooperating/mentor teachers. Just as the essential feature of classroom inquiry varies along a continuum, the preservice and in-service teachers' development and professional needs along the continuum vary as well. In order to reach all students, it is imperative that teachers are able to enact promising practices such as inquiry-based learning so that all students have the opportunity to achieve scientific literacy and ensure a more equitable education (Darling-Hammond, 2010; Ravitch, 2010). Through ongoing coaching/mentoring in a variety of situations both on/off campus, the journey of inquiry into inquiry can continue to evolve for preservice teachers becoming increasingly generative and invitational as both time and situation unfold infinitely into the future.

The physics inquiry course has been incorporated into the degree plans of those who intend to graduate with either a BS/BA in Physics with a *teach*HOUSTON minor degree or a Mathematics or Science degree with a physics minor.

Additionally, this course was added to the degree plans of those seeking science composite certification as many end up teaching physics. Jason's experience and comments on the post survey were used to influence the direction of the curriculum for the physics inquiry course, i.e., more time is devoted to discussing how to enact inquiry-based learning in high schools and potential roadblocks and overcoming these roadblocks. Because of these responses along with the feedback of others, the physics inquiry course now incorporates discussion of roadblocks, and also a section over lesson planning with inquiry-based learning. This research has the potential to inform other teacher preparation programs to include courses whereby physics content, among other science subjects, is taught as inquiry.

AUTHORS' NOTE

This research was funded by the following grants:

- National Science Foundation DUE Award 1240083, *Recruitment, Preparation and Retention of STEM Students as High School Teachers*
- National Science Foundation DUE Award 1557309, *University of Houston: Learning through Informal and Formal Experiences*
- Department of Education Supporting Effective Educator Development (SEED) Grant, U423A180139, *Expanding and Strengthening the STEM Teacher Workforce Through UTeach*

REFERENCES

American Association for the Advancement of Science. (1993). *Benchmarks for science literacy: A project 2061 report*. New York, NY: Oxford University Press.

Anderson, R. D., & Mitchener, C. P. (1994). Research on science teacher education. In D. L. Gable (Ed.), *Handbook of research on science teaching and learning* (pp. 3–44). New York, NY: Macmillan.

Azam, S. (2018). Physics for teaching high school physics: Views of prospective physics teachers and teacher educators about undergraduate physics study. *Journal of Teacher Education and Educators*, *7*(2), 147–163.

Basit, T. N., & McNamara, O. (2004). Equal opportunity or affirmative action? The induction of minority ethnic teachers. *Journal of Education for Teaching: International Research and Pedagogy*, *30*, 97–115.

Clandinin, D. J. (1986). *Classroom practice: Teacher images in action*. London: Falmer Press.

Clandinin, D. J., & Connelly, F. M. (1992). Teacher as curriculum maker. In P. W. Jackson (Ed.), *Handbook of research on curriculum* (pp. 363–401). New York, NY: Macmillan.

Clandinin, D. J., & Connelly, F. M. (2000). *Narrative inquiry: Experience and story in qualitative research*. San Francisco, CA: Jossey-Bass.

Connelly, F. M., & Clandinin, D. J. (1990). Stories of experience and narrative inquiry. *Educational Researcher*, *19*(5), 2–14.

Craig, C. (2009). The contested classroom space: A decade of lived education policy in Texas schools. *American Educational Research Journal*, *46*(4), 1034–1059.

Craig, C. (2010a). Coming full circle: From teacher reflection to classroom action and places in-between. *Teachers and Teaching: Theory and Practice*, *16*(4), 423–435.

Craig, C. (2010b). Teacher as curriculum maker. In C. Kridel (Ed.). *The encyclopedia of curriculum studies* (Vol. 2, pp. 867–869). Thousand Oaks, CA: Sage Publications.

Craig, C. (2011). Teacher professional development through a teacher-as-curriculum maker lens. In M. Kooy, & K. van Veen (Eds.), *Teacher learning that matters*. Singapore: Springer.

Craig, C., & Ross, V. (2008). Cultivating teachers as curriculum makers. In F. M. Connelly (Ed.), *Sage handbook of curriculum and instruction* (pp. 282–305). Thousand Oaks, CA: Sage Publications.

Darling-Hammond, L. (1997). *Doing what matters most: Investing in quality teaching*. New York, NY: National Commission on Teaching & America's Future.

Darling-Hammond, L. (2010). *The flat world and education: How America's commitment to equity will determine our future*. New York, NY: Teachers College Press.

Darling-Hammond, L. (2012). *Powerful teacher education: Lessons from exemplary programs*. Hoboken, NJ: John Wiley & Sons.

Dee, T. S. (2005). A teacher like me: Does race, ethnicity, or gender matter? *The American Economic Review*, *95*, 158–165.

Dewey, J. (1897, January 16). My pedagogical creed. *The School Journal*, *54*(3), 77–80.

Dewey, J. (1933). *How we think*. Chicago, IL: Henry Regnery.

Dewey, J. (1938). *Experience and education*. New York, NY: Macmillan.

Dewey, J. (1944). *Democracy and education: An introduction to the philosophy of education*. New York, NY: Free Press.

Dubson, M. (2007). Three or four golden rules of lecture. *The Physics Teacher*, *45*(4), 252–253.

Feiman-Nemser, S. (1983). Learning to teach. In L. S. Shulman, & G. Sykes (Eds.), *Handbook of teaching and policy* (pp. 150–170). New York, NY: Longman.

Fitzgerald, M., Danaia, L., & McKinnon, D. H. (2019). Barriers inhibiting inquiry-based science teaching and potential solutions: Perceptions of positively inclined early adopters. *Research in Science Education*, *49*, 543–566.

Goodrum, D., & Rennie, L. J. (2007). *Australian school science education national action plan, 2008-2012*. Canberra: Department of Education, Science and Training.

Jasien, P. G., & Oberem, G. E. (2002). Understanding of elementary concepts in heat and temperature among college students and K-12 teachers. *Journal of Chemical Education*, *79*(7), 889.

Kliebard, H. M. (2004). *The struggle for the American curriculum*. London: Routledge Falmer.

Lawson, A. E. (2002). The learning cycle. In R. G. Fuller (Ed.), *A love of discovery: Science education, the second career of Robert Karplus* (pp. 51–62). New York, NY: Kluwer Academic.

Lederman, N. G., Gess-Newsone, J., & Latz, M. S. (1994). The nature and development of pre-service science teachers' conceptions of subject matter and pedagogy. *Journal of Research in Science Teaching*, *31*(2), 129–146.

MacLeod, K. (2013). Physics education and STSE: Perspectives from the literature. *European Journal of Physics Education*, *4*(4), 1–12.

Marshall, J. C., Smart, J. B., & Alston, D. M. (2016). Inquiry-based instruction: A possible solution to improving student learning of both science concepts and scientific practices. *International Journal of Science and Mathematics Education*, *15*(5), 777–796.

McDermott, L. C. (1984). Research on conceptual understanding in mechanics. *Physics Today*, *37*(7), 24–32.

McDermott, L. C. (1990). A perspective on teacher preparation in physics – other sciences: The need for special courses for teachers. *American Journal of Physics*, *58*(8), 734–742.

McDermott, L. C. (1993). Guest comment: How we teach and how students learn—A mismatch? *American Journal of Physics*, *61*, 295–298.

McDermott, L. (1996). *Physics by inquiry: Volume I and Volume 2*. New York, NY: Wiley.

McDermott, L. (2007). Preparing K-12 teachers in physics: Insights from history, experience, and research. *American Journal of Physics*, *74*, 758–762.

McDermott, L. C., Shaffer, P. S., & Constantinou, C. P. (2000). Preparing teachers to teach physics and physical science by inquiry. *Physics Education*, *35*(6), 411–416.

McDermott, L. D., & Shaffer, P. S. (1992). Research as a guide for curriculum development: An example from introductory electricity. Part 1 Investigation of student understanding. *American Journal of Physics*, *60*(11), 994–1003.

McNeil, L., & Valenzuela, A. (2001). The harmful impact of the TAAS system of testing in Texas: Beneath the accountability rhetoric. In M. Kornhaber & G. Orfield (Eds.), *Raising standards or raising barriers? Inequality and high stakes testing in public education* (pp. 127–150). New York, NY: Century Foundation.

Meltzer, D. E., & Thornton, R. K. (2012). Resource letter ALIP–1: Active-learning instruction in physics. *American Journal of Physics*, *80*(6), 478–496.

National Academy of Sciences. (2007). *Rising above the gathering storm: Energizing and employing America for a brighter economic future*. Washington, DC: National Academy Press.

National Center for Education Statistics. (2018). Characteristics of public school teachers. Retrieved from https://nces.ed.gov/programs/coe/pdf/Indicator_CLR/coe_clr_2018_05.pdf.

National Research Council. (1996). *National science education standards*. Washington, DC: National Academy Press.

National Research Council. (2000). *Inquiry and the national science education standards: A guide for teaching and learning*. Washington, DC: National Academy Press.

National Research Council. (2005). *America's lab report: Investigations in high school science*. Washington, DC: National Academy Press.

National Research Council. (2011). *Assessing 21st century skills: Summary of a workshop*. Washington, DC: National Academies Press. doi:10.17226/13215.

National Research Council. (2012). *Education for life and work: Developing transferable knowledge and skills in the 21st century*. Washington, DC: National Academies Press.

Nichols, S. L., & Berliner, D. C. (2005). *The inevitable corruption of indicators through high stakes testing*. Tempe, AZ: Arizona State University, Educational Policies Research Unit.

Olson, M. R. (1995). Conceptualizing narrative authority: Implications for teacher education. *Teaching and Teacher Education*, *11*(2), 119–135.

Pereira, P. (2005). Becoming a teacher of mathematics. *Studying Teacher Education*, *1*(1), 69–83.

Pitts, D. W. (2007). Representative bureaucracy, ethnicity, and public schools: Examining the link between representation and performance. *Administration & Society*, *39*, 497–526.

Polkinghorne, D. (1995). Narrative configuration in qualitative analysis. In J. A. Hatch, & R. Winiewski (Eds.), *Life history and narrative* (pp. 5–23). Washington, DC: The Falmer Press.

President's Council of Advisors on Science and Technology. (2010). *Prepare and Inspire: K-12 Education in science, technology, engineering, and math (STEM) for America's future*. Washington, DC: Executive Office of the White House.

Putnam, R. T., & Borko, H. (1997). Teacher learning: Implications of new views of cognition. In B. J. Biddle, T. L. Good, & I. F. Goodson (Eds.), *International hand-book of teachers & teaching* (Vol. II, pp. 1223–1296). Dordrecht: Kluwer.

Ramsay, M. C. (2019). SBEC online data, TEA PEIMS data. Retrieved from https://tea.texas.gov/sites/default/files/Certified%20Teachers%20by%20Preparation%20Route%202014-2018.pdf.

Ravitch, D. (2010). *The death and life of the great American school system: How testing and choice are undermining education*. Basic Books.

Razali, S. N., & Yager, R. E. (1994). What college chemistry instructors and high school chemistry teachers perceive as important for incoming college students? *Journal of Research in Science Teaching*, *31*(7), 735–747.

Redish, E. F., & Steinberg, R. N. (1999). Teaching physics: Figuring out what works. *Physics Today*, *52*, 24–30.

Sadler, P. M., & Tai, R. H. (2001). Success in college physics: The role of high school preparation. *Science Education*, *85*(2), 111–136.

Schwab, J. (1960a). Enquiry, the science teaching and the educator. *The Science Teacher*, *27*(6), 6–11.

Schwab, J. J. (1960b/1978). The practical: Translation into curriculum. In I. Westbury, & N. Wilkof (Eds.), *Science curriculum and liberal education: Selected essays* (pp. 365–383). Chicago, IL: University of Chicago Press.

Schwab, J. (1962). The teaching of science as enquiry. In J. Schwab, & P. Brandwein (Eds.), *The teaching of science* (pp. 3–103). Cambridge, MA: Harvard University Press.

Schwab, J. J. (1983). The practical 4: Something for curriculum professors to do. *Curriculum Inquiry*, *13*(3), 239–265.

Silm, G., Tiitsaar, K., Pedaste, M., Zacharia, Z. C., & Papaevripidou, M. (2017). Teachers' readiness to use inquiry-based learning: An investigation of teachers' sense of efficacy and attitudes toward inquiry-based learning. *Science Education International*, *28*(4), 315–325.

*teach*HOUSTON Program. (2017). Description of program [About]. Retrieved from http://teach-houston.uh.edu/.

Texas Education Agency. (2017). Out of field teaching report, 2016-2017. Retrieved from https://tea.texas.gov/Reports_and_Data/Educator_Data/Educator_Reports_and_Data/.

Texas Education Agency. (2018). Teacher retention 2014-2018. Retrieved from https://tea.texas.gov/sites/default/files/Teacher%20Retention%20by%20Route%202014-2018.pdf

Townsend, A., McKinnon, D., Fitzgerald, M., Morris, J., & Lummis, G. (2017). Educative curricula and PCK development driven by STEM professional learning in rural and remote schools: A longitudinal type IV case study. *International Journal of Innovative Science and Modern Engineering*, *24*(4), 1–17.

UTeach website. (2017). Description of program [About]. Retrieved from http://www.uteach-institute.org/.

Wainwright, C. (Ed.). (2006, March 24–25). The CASTLE curriculum: Capacitor-aided system for teaching and learning electricity. Paper presented at 2006 PhysTEC Conference, Fayetteville, Arkansas. Retrieved from https://www.compadre.org/Repository/document/ServeFile.cfm?ID=4438&DocID=164.

Windschitl, M. (2002). Framing constructivism in practice as the negotiation of dilemmas: An analysis of the conceptual, pedagogical, cultural, and political challenges facing teachers. *Review of Educational Research*, *72*(2), 131–175.

ENHANCING PRESERVICE TEACHER PREPARATION THROUGH FORMAL AND INFORMAL LEARNING EXPERIENCES

Donna W. Stokes and Paige K. Evans

ABSTRACT

Learning through formal and informal experiences is critical for building content knowledge, pedagogical skills, and self-efficacy/confidence for preservice teachers. teachHOUSTON offers numerous teacher enhancement opportunities outside the teacher education courses which allows preservice teachers to connect to the real world which includes being able to relate to a diverse population of students and to understand how the course content can be related to them, their families, and communities in their everyday experiences. Through formal and informal experiences such as professional development workshops, discipline specific courses, research experiences, and internships, preservice teachers have the opportunity to engage in hands-on science activities they can use with their students, develop lessons, and gain knowledge on how to deliver this content while managing their classroom. This chapter will give an overview of the formal and informal experiences offered through teachHOUSTON with a highlight on the structure and content of the six week Noyce Internship Program which engaged interns as counselors and teaching assistants in a summer STEM camp for underserved middle school students and introduces the interns to interactive sessions that model promising practices for teaching.

Keywords: Formal learning; informal learning; summer internship; STEM; teacher education; professional development; self-efficacy; *teach*HOUSTON

Preparing Teachers to Teach the STEM Disciplines in America's Urban Schools
Advances in Research on Teaching, Volume 35, 65–83

ISSN: 1479-3687/doi:10.1108/S1479-368720210000035005

ENHANCING PRESERVICE TEACHER PREPARATION THROUGH FORMAL AND INFORMAL LEARNING EXPERIENCES

Providing students with a solid foundation in Science, Technology, Engineering and Mathematics (STEM) areas is important in the education of future scientists, doctors, and engineers. Important in this preparation are STEM teachers at all levels of the education system, i.e., secondary and postsecondary, who can educate and inspire students to pursue STEM careers. Expert teachers and experienced leaders are essential to improving STEM education which requires substantial investments in teacher quality (College Board, 2008). Thus, it is essential to work toward helping educators embrace learner-focused methods that prepare students to become productive members of society (Darling-Hammond, 2006). State and federal education authorities and industry leadership demand improvement in the preparation of math and science teachers (Zientek & Thompson, 2008).

Attracting highly qualified teachers to teach in high needs schools is challenging as many schools lack adequate support and funding/resources. The education of students in the United States has been a growing concern for our nation since the *A Nation at Risk* (1983) report which assessed the quality of teaching and learning in US schools, i.e., public/private schools, colleges, and universities. Many have championed that teacher education may be a solution to meeting the challenges faced by the United States to meet the supply and demand for teachers (Gardner, 1983). In STEM fields, training of quality teachers is challenging as it brings additional layers to unfold as many STEM teacher education programs are housed in education departments while others are housed in STEM departments. In addition, the lack of widespread agreement on who, how, and what types of teacher education models are best for training qualified STEM teachers has been problematic. To this end, how to educate students to teach in these disciplines has been an area of research for many teacher education researchers. Their research includes identifying promising practices for student learning/teaching, determining effective methods to building a strong STEM content knowledge, understanding how to engage students considering cultural contexts to relate to students in high needs schools, and choosing the proper settings, formal and informal, to best prepare and train STEM teachers.

LEARNING THROUGH FORMAL AND INFORMAL ACTIVITIES

Learning in both formal and informal settings (Aslam, Adefila, & Bagiya, 2018; Kim & Keyhani, 2019) in teacher preparation programs builds content knowledge, pedagogical skills, and self-efficacy. Learning in a formal setting, i.e., traditional classroom settings, professional development workshops, conferences sessions, and lab-based research experiences, is important, but informal experiences, which can be gained through clinical practices, are also essential for building student self-efficacy to be an effective teacher. In particular, skills gained through internships, which include both formal and informal activities, engage students in real-world, field-based experiences and allow them to use their personal lens to see how activities outside the classroom can result in teaching and learning.

Internships have been defined in several ways as it applies across disciplines (O'Neill, 2010). The National Association of Colleges and Employers (2011) defines internships as "...a form of experiential learning that integrates knowledge and theory learned in the classroom with practical application and skills development in a professional setting."

Dean (2006) states:

> The primary mission of Internship Programs is to engage students in planned, educationally related work and learning experiences that integrate knowledge and theory with practical application and skill development in a professional setting. (Council for the Advancement of Standards in Higher Education, 2006, p. 4).

In reviewing the plethora of internship definitions, common terminology of the definitions include experiential learning, field-based experiences, academic experience which integrates theory with practice, professional connection, and active learning in a professional setting to name a few. One definition as it relates to teacher education states that internships are:

> ...extended field based and context-responsive professional learning experiences negotiated collaboratively by stakeholders in the culminating phase of preservice teacher preparation... The intern is mentored and immersed in a broad range of teachers' professional work activities. It involves a shift in status for the preservice teacher with increased opportunities for autonomy, responsibility and accountability but with a safety net. (Board of Teacher Registration, Queensland, 2003, p. 7).

Although this definition is fitting for a variety of teacher education internships, the actual definition of, and experience gained, will be determined by the host of the internship. Therefore, internships in teacher education programs can look very different from program to program with each having a different goal. They can range from long-term formal opportunities for building knowledge, skills, and competencies to short-term training activities in informal settings for utilizing and assessing skills and knowledge gained in the teacher education classes before entering the teaching profession.

Internships can engage students individually or in teams for building teacher identity and/or self-efficacy. They can also serve as recruitment/retention tools and serve as a self-assessment and improvement platform. Internships in formal settings are often required parts of teacher education programs while those in informal settings may or may not be. It is important for preservice teachers to include training opportunities such as internships which include both formal and informal experiences to transition preservice teachers from being a student learner to a teacher of students.

BEYOND TRADITIONAL TEACHER TRAINING: ADDITIONAL OPPORTUNITIES THROUGH *teach*HOUSTON

In order for teacher education programs to prepare effective STEM teachers, they must offer additional teacher training opportunities besides the traditional teacher education courses required for STEM teacher certification. The structure of additional teacher training within teacher education programs is a layered process which should strengthen content knowledge and build teaching skills through professional development. Approaches to teaching science and how to connect it to the real world should be an ongoing process as teachers need to be able to relate the content to their students as it pertains to them and their families in modern times. This includes being able to relate to a diverse population of students and to understand how the course content can be related to their everyday experiences. In formal teacher education classes, preservice teachers may or may not have the opportunity to engage in hands-on science activities they can use with their students, develop lessons, and get feedback from their peers, or have the opportunity to gain knowledge on how to deliver this content while managing their classroom.

PROFESSIONAL DEVELOPMENT

Professional development is often used to allow preservice teachers the opportunity to obtain exposure to various aspects of teaching. However, professional development is often delivered in similar ways that science and teacher education courses are taught, in a traditional, formal lecture style covering the science and math content/pedagogy. Research shows that informal STEM settings can be utilized to deliver professional development to address these challenges and allow in- and preservice teachers to solidify their content knowledge, build their self-efficacy in applying teaching techniques, and connect the science to the real world by emphasizing the use of critical thinking and reasoning skills all through interactive learning/teaching environment (Jackson & Mohr-Schroeder, 2018; Terrazas-Marín, 2018).

Researchers have defined professional development standards for science teachers which consider not only the nature of the professional development but also the context for which it takes place (National Research Council, 1996). These standards require that science teacher professional development: (1) is a continuous process throughout the teachers career; (2) includes training opportunities which stimulate intellectual professional growth; (3) connects to the teachers' role in the context of the school and (4) allows teachers to grow through experiences which will allow them to help their students. These standards can be met in a number of ways, but meeting these standards is challenging for those providing/seeking science teacher professional development. How do in- or preservice teachers identify professional development opportunities and how do teacher education programs offer experiences which will allow them to meet these standards? Teacher education programs should take on the responsibility of offering professional development which can engage its preservice teachers as well as graduates/in-service teachers from its program in meaningful professional development activities. Oftentimes, particularly with high needs schools, professional development is limited due to budgetary constraints; therefore, teachers are left with the responsibility of having to seek funding or provide their own funding for ongoing professional development which will allow them to maintain the knowledge/skills needed to keep up with teaching techniques and with ever changing technology for content delivery. Hence, professional development must occur within the teacher education program's formal and informal activities (Du et al., 2019; Stokes, Evans, & Craig, 2017). *teach*HOUSTON does this in a variety of ways including workshops, courses, conferences/meetings, research experiences, internships, community service events, and networking activities, which are outlined below.

TRADITIONAL WORKSHOPS

*teach*HOUSTON offers formal professional development through approximately 20 professional development workshops which cover topics relevant for preservice teachers in the fall and spring semesters. Several examples of the workshops covered are shown in Table 1.

These workshops offer the students various opportunities to supplement what they are learning in their *teach*HOUSTON courses. Topics included in the professional development series are suggestions by current faculty in the *teach*HOUSTON program and also requested by current students. Workshops are conducted by current and former students as well as faculty in the *teach*HOUSTON program. For example, students in the field-based courses (see Chapter 2), utilize the 5E lesson plan template to create engaging and student-centered lessons to teach in high-need school districts, where the 5E constructivist

Table 1. Professional Development Workshop Topics for a Typical Fall and Spring Semesters.

Workshop Topics	
Fall Semester	Spring Semester
PPR 101	*teach*HOUSTON Ambassador
5 E's and Me	Acing the Apps
The Clinical Interview	Managing Microsoft Office
Managing a Caring Classroom	Super Saturday Professional Development Session
Equipment (Probes)	Beyond the Classroom: Careers in Education
It Takes a Village	Teaching English Language Learners
Closing the Achievement Gap	Breaking STEM Gender Stereotypes in the Classroom
Crushing the Content	Pedagogy of Poverty
Teaching LGBTQ + Youth	Public Education Advocacy
Socratic Teaching	Stopping the School to Prison Pipeline
Navigating Through *teach*HOUSTON	Supporting Undocumented Students and Mixed Status Families
Technology Integration	

model includes the five phases of learning/teaching to promote student inquiry: engage, explore, explain, elaborate, and evaluate. However, many students need additional support when creating their lessons. Thus, the *5E's and Me Professional Development* session was created and goes over tips and tricks for creating highly effective 5E lesson plans as well as providing resources for ideas and tips for each E as well as for lesson plans as a whole. Additionally, students are required to conduct a clinical interview early on in one of the teacher certification classes. The *Clinical Interview Professional Development* discusses the importance of conducting clinical interviews in class to ensure that students understand the material, and it explains the process of conducting the interviews and how to use them to improve lessons. Another workshop topic that is of interest to the majority of the *teach*HOUSTON students is classroom management. The *Managing a Caring Classroom Professional Development* discusses how to manage a classroom in a way that establishes authority, while also creating a productive, nurturing learning environment. From classroom arrangement to teacher–student interactions, this professional development discusses the best ways to achieve success in the classroom. In the Research Methods class, students are required to use Probeware for collecting and analyzing data. Thus, *the Classroom Equipment (Probes) Professional Development* explains how to use the technology in the classroom, specifically in science classrooms. It discusses how and when to use probes and how students benefit from using technology in and outside the classroom. In addition to those listed in Table 1, other frequently requested professional development workshops on how to successfully pass the teaching certification exams, how to manage time, bullying in the classroom, and collaboration in the classroom to name a few.

DISCIPLINE SPECIFIC TEACHER EDUCATION COURSES

Formal activities of courses offered through *teach*HOUSTON also engage the students in professional development. For example, a Physics by Inquiry course was developed and implemented into the *teach*HOUSTON program as part of the University of Houston's National Science Foundation Noyce Scholarship Program, which offers scholarships and internships to recruit and retain physics and chemistry majors to become secondary teachers. The formal activities associated with the course allows Noyce scholars as well as other *teach*HOUSTON students to learn physics through inquiry and to reinforce what was learned in the physics course which is a part of their major degree plan. The major courses focus on teaching the physics content, but often are taught in a traditional style that does not provide a model for teaching science at the K–12 level. McDermott, Shaffer, and Constantinou (2000) stated as it relates to physics instruction, "Teachers tend to teach as they were taught. If they were taught through lecture, they are likely to lecture, even if such instruction is inappropriate for their students" (p. 412).

By learning physics through inquiry in the Physics by Inquiry course, preservice teachers are exposed to promising practices for learning/teaching physics which allows them to learn physics through inquiry while using critical thinking and reasoning skills, and allows them to develop lessons which they can use in their own classrooms. The experiences gained through the formal classroom-based activities of the Physics by Inquiry course help to build self-confidence for the preservice teachers as they strengthen their physics content knowledge. These activities allow the teachers to reflect on their own learning process and how it can be used in their future classes. The insights gained by learning the content in the way they will teach it to their students put the teaching/learning techniques in context as they apply to the science content. This "built in" professional development through the formal context of the course emphasizes the scientific concept needed for building self-efficacy of the preservice teachers.

CONFERENCE/MEETING ATTENDANCE

*teach*HOUSTON students also gain formal and informal professional development through conference and meeting attendance. *teach*HOUSTON students attend annual Noyce Regional, UTeach, and Houston Areas Student Teaching Conferences, as well as the Conference for Advancement of Science Teachers, Texas Section of the American Physical Society, and the Association of Physics Teachers meetings. The activities at conferences such as these have been shown to have positive outcomes on teacher training and professional development (Hunzicker, Biddison, Hickey, & McClethen, 2014). For example, 16 *teach*HOUSTON preservice and in-service teachers who were Noyce scholars expressed their appreciation to the organizers of the Western Regional Noyce conference they attended in 2017 in Fresno, CA (as mentioned in Chapter 3). Student responses outlined below demonstrated the personal and professional benefit of attending the conference. Increases in self-efficacy and motivation were gained as Student 1 stated:

> Thank you for the opportunity to interact with like-minded people. This conference provided me with an incredible opportunity to grow as a teacher and to passionately continue this incredible journey. After this conference, I commit myself to continue working hard for my students every day.

The conference workshops/session (attended and presented) offered the students the opportunity to engage in professional development which would be useful as a pre- or in-service teacher. Student 2 stated:

> [The] WRNC has provided me with invaluable information obtained from its workshops and has helped me develop professionally by allowing my group and I to present at the conference. I am incredibly grateful for this wonderful experience and would like to thank everyone for this opportunity.

In addition, networking with other preservice and in-service teachers at the conference helped the students build up a support system which included information about teacher resources. Student 3 stated:

> The WRNC helped me connect with other teachers from around the country. My list of resources grew as well as my available opportunities.

Attending workshops and presenting at conferences is an important part of teacher education as it engages the students with pre-and in-service teachers outside of their program, provides networking to gain classrooms and personal resources for becoming a great teacher. One of the great aspects of the Western Regional Noyce Conference is that it fully supports (housing, meals, and travel) the Noyce Scholar offering them free professional development which may not be possible for these students once they become teachers in high-need schools.

RESEARCH EXPERIENCE

*teach*HOUSTON students also have opportunities to participate in formal on- and off-campus lab research programs. Many teacher education programs integrate science-based lab research experiences in programs as a method of professional development for building skills needed for conducting/teaching lab procedures in the classroom. Research on the effectiveness of a laboratory research experience as a tool for practical training of teachers for content delivery in the classrooms for secondary teachers has shown that preservice teachers gain a direct connection and appreciation for the "science" and how it can be related to the content they will be teaching (Amolins, Ezrailson, Pearce, Elliott, & Vitiello, 2015; Westerlund, Schwartz, Lederman, & Koke, 2001). Other research shows that by utilizing a constructivist model, where preservice teachers conduct science research with scientist in the field, it enables them to learn in ways that will allow them to transfer their knowledge to the students in their classroom based on their experience (Brown, Bolton, Chadwell, & Melear, 2002; Melear, Goodlaxson, Warne, & Hickok, 2000).

Learning through research-based experiences for teacher training has been supported by the National Science Foundation as it sponsors many programs including the Research Experiences for Teachers (RET) (Pop, Dixon, & Grove, 2010; Russell & Hancock, 2007a, 2007b; Southerland et al., 2016; Wilhelm & Fisher, 2019). The success of the RET Program, which aims to improve science education by exposing teachers to research, has been well documented (McCarthy, 2011; Pozo, Martinez-Aznar, Rodrigo, & Varela, 2004; Wenglinsky & Silverstein, 2007). While this type of teacher training has been shown to be successful in the development of laboratory skills, translation into the classroom has not been as successful. In light of this, other research-based programs for teacher training have been modeled after the RET program, but these include some instructional and pedagogical professional development to promote and assist with translation of the research gained knowledge into the classroom. The Science Teacher and Researcher (STAR) program, which is a partnership between California Polytechnic State University (Cal Poly), K–12 districts, federal research agencies including National Science Foundation (NSF), and industry, also aims to train science and math teachers through lab-based research, is an example of a program that includes professional development for connecting the science curriculum/content to knowledge gained from the lab research to inspire students to pursue STEM careers (Baker & Keller, 2010).

*teach*HOUSTON students have not only participated in research in labs with faculty in the physics and chemistry Departments at University of Houston (UH), but also in research through the STAR Program (Baker & Keller, 2010). The STAR Program offers a nine-week summer fellowship to Noyce scholars in teacher education programs to gain hands-on research experience at research centers across the United States. Fellows work on formal research projects under the supervision of a mentor, attend professional development workshops aimed to assist them with translating their research into inquiry-based lessons, build a network with preservice teachers from all over the nation and present their research results at the Program's Closing Conference. Participating labs include Brookhaven National Laboratory, NASA Jet Propulsion Lab, NASA Armstrong Flight Research Center, and SLAC National Accelerator Laboratory. As Student 4, a participant in the STAR program at multiple sites stated:

> As a *teach*HOUSTON alumnus and Noyce Graduate, I have had many opportunities to learn and growth as a student, a researcher, and as a teacher by attending many STEM education conferences and participating in [research at] many prestigious US national laboratories such as SLAC, BNL, PNNL, and Fermilab. These amazing opportunities have a direct benefit to my students from tailored inquiry-based lessons that relate physics concepts to cutting-edge scientific research at US national laboratories, as well as, accelerate and diversify my career as a STEM educator.

This shows that engagement in lab research allows preservice teachers the opportunity to expose their students to cutting edge research in STEM fields using the knowledge they gained through their experience. This can result in improved preservice teacher confidence, better connection of the content to real science problems which may be related to their student's communities and improved ability for the preservice teacher to promote STEM careers to their students.

INTERNSHIPS

Internships can encompass all the previously discussed training outlets for preservice teachers. Through *teach*HOUSTON, a variety of internships are available which offer training and professional development that allow preservice teachers to connect through individual or group formal and informal activities to meet their teaching needs. This includes student worker internships where preservice teachers work for *teach*HOUSTON Master Teachers and internships as facilitators/counselors for summer camps. Working with Master Teachers, the students are able to work one-on-one with faculty in preparing course materials and lessons in the introductory courses in the *teach*HOUSTON program. This gives preservice teachers insight as to what is involved in creating and delivering lessons. Additionally, these interns also serve as mentors for those in introductory classes. They are able to work with students on developing and practicing lesson plans, for example the 5E lesson plan, and they will eventually teach in high-need school districts. These internships allow preservice teachers to gain valuable experience in mentoring where they gain insight into personal and professional growth while also honing their lesson planning skills as they are providing resources and advice to their mentees.

NOYCE INTERNSHIP PROGRAM

One of the most prestigious internship opportunities offered through *teach*HOUSTON is the Noyce Internship Program which has training through teaching experiences, curriculum, and professional development. As part of the NSF Robert Noyce Scholarship Program, the Noyce Internship Program was developed and implemented to provide formal and informal activities where it: (1) serves as a recruitment/retention tool for physics/chemistry majors to the *teach*HOUSTON program, (2) exposes students to teaching experiences early in their academic career, (3) provides informal experiences for teaching/learning, (4) offers formal hands-on science experimentation and lesson development, and (5) provides professional development on relevant topics.

The four week summer Noyce Internship Program is open to freshmen NSM majors who have completed the introductory, STEP 1 course (see Chapter 2) which offers STEM majors the opportunities to "try out" teaching where students are engaged in classroom teaching in their first semester giving them a taste of what teaching requires. The students in this course are encouraged to participate in an internship experience to solidify their choice to pursue a teaching career.

STEM SUMMER CAMP

The Noyce Internship Program utilized a STEM camp to offer both informal and formal experiences and professional development for training preservice teachers. Interns served as counselors in one or more of the summer camps offered through *teach*HOUSTON. The two-week camps serviced students who attended high needs schools; therefore, the interns worked with the population they will be teaching. This included: (1) the Exxon Mobil Bernard Harris Summer Science Camp, a residential camp which provided activities, experiments, projects, and field trips for underserved students entering 6th–8th grade, (2) the Chiyoda Young Innovators' Academy that exposed 7th–9th grade students to engineering, programming and inventions, and/or (3) the StatOil STEM Camp where 5th–8th graders engage in themed integrated science, technology, engineering, and math activities. These camps aimed to increase campers' STEM skills and stimulate their interest in STEM careers through math and science inquiry based lessons developed by the interns.

Noyce interns interacted with campers daily by assisting Master Teachers with instruction and serving as camp counselors. They were expected to have conceptual understanding and experiential learning to develop inquiry and quantitative based science curriculum and be equipped to use various approaches for teaching the campers. The interns were also expected to know policies/procedures, which govern the interns as well as the campers, and possess the skills necessary to be effective camp facilitators. The skill set required dictated that they should be able to:

(1) Recognize signs of child abuse and understand the reporting procedures that accompany it;
(2) Know and state the dress code that is appropriate for summer camps;
(3) Explain the counselor responsibilities and expectations;
(4) Demonstrate different forms of professionalism;
(5) Execute effective facilitation skills in small group settings;
(6) Distinguish between different types of middle schoolers;
(7) Implement college and career readiness strategies in the summer camp;
(8) Utilize classroom management skills throughout the summer camp;
(9) Use the procedures in the safety plan to follow safety guidelines for the summer camp;
(10) Review and improve the lessons and make suggestions for future camps to ensure the interns possess the skills needed to work during the camps and beyond, professional development was offered through and within the summer STEM camps.

NOYCE INTERNSHIP INSTITUTE: PROFESSIONAL DEVELOPMENT

The Noyce Internship Program also offered interactive professional development sessions through the Noyce Internship Institute which aimed to address preservice teacher needs as it relates to developing/implementing math and science lessons and for developing tools/professional skills necessary for teaching. Research shows that professional development, individually or in groups, aimed toward developing and testing curriculum helps to build teacher knowledge and confidence on how to tailor their instructional materials to connect with the students based to the environment available for teaching such as classrooms in high needs schools which may not have the most up-to-date technology/equipment (Ball & Cohen, 1996; Coenders, Terlouw, Dijkstra, Pieters, & Pieter, 2010; McFadden & Roehrig, 2017). The two-week institute occurred before the start of the summer camps. It offered approximately 60 hours of professional development activities for improving content knowledge and instructional approaches/strategies for teaching by utilizing promising practices, i.e., teaching tools and techniques and strategies for improving student outcomes. The sessions were delivered by *teach*HOUSTON Master Teachers, NSM faculty, and previous interns with leadership capacity to demonstrate strategic use of teaching tools to promote inquiry- and project-based learning and to ensure accurate delivery of the science content.

The Noyce Internship Institute ran for 6 hours each day, with time dedicated to: (1) professional development in content specific areas including inquiry-based lesson/curriculum development and testing; (2) implementation of instructional strategies and relevant topics such as project-based learning; (3) use of technology in the classroom as this is a challenge for teachers as the field is always changing (Waight & Abd-El-Khalick, 2012); (4) classroom management, (5) introduction of teaching strategies based on promising practices; and (6) workshops related to working in the STEM camp as well as in their future classrooms. Inquiry-based lessons in STEM were developed, modified, and refined for use in the classroom. In addition, training on topics such as Cultural Competence, Professionalism and Growth Mindset, Building Relationships, Arduinos, Ice-breakers for Middle School Students, Classroom Management, Culturally Responsive Pedagogy (CRP), and Maker Space Activities are included. Training guides for the two-week professional development, shown in Tables 2 and 3, were developed and can be used as a model for other teacher education programs. A detailed outline of the activities, workshops and content-focused sessions are shown. The institute encouraged group work which will be expected as a preservice teacher's transition to working in difficult-to-staff secondary schools. For example, each day, the interns engaged in a team building activity designed to foster cohesiveness and teamwork which will be essential when serving as the teaching/counseling team in the camp. Further to this, it is hoped that the importance of teamwork and collegiality will extend beyond the internship into their professional careers when working in professional learning communities as most content teams in schools plan their curricula together as professional learning communities. According to Hord (1997), professional learning communities should encourage constructivism in that the self-initiated learner works with peers to improve teaching quality. Thus, it is expected that preservice teachers will be more confident in planning with future colleagues as a result of this experience.

The institute also focused on training students to be culturally conscious as they will be serving as counselors for STEM camps and ultimately teaching in high needs schools which serve students from underrepresented groups. CRP, which has been incorporated into the *teach*HOUSTON teacher certification courses, was also embedded in the Noyce Internship Institute. CRP as defined by Gay (2010) entails

> …using the cultural knowledge, prior experiences, frames of reference, and performance styles of ethnically diverse students to make learning encounters more salient to them. Furthermore, it teaches to and through the strengths of these students.

It is important that students be trained in ways that include a cultural context in STEM classrooms for a number of reasons. In the next decade, about 80% of the jobs will be related to STEM fields. However, about 50% of students with a STEM major do not complete the degree particularly underrepresented minorities. For example, Houston is one of the most diverse cities in the nation, and Colby and Ortman (2014) project that by 2044, over half of all Americans will belong to a minority group. Therefore, it will be important that educators determine and utilize promising practices for training all STEM students, particularly teachers who will train the future STEM workforce. Defining and understanding culture is very complex, and it becomes even more complex when it has to be addressed in the context of STEM teaching/learning; therefore, preservice STEM teachers must understand how to proceed to include/consider culture, in its complexity, in classrooms of very diverse students.

Table 2. NOYCE Intern Week One Training Overview.

NOYCE Intern Week One Training Overview				Sponsored by:	teachHOUSTON
	Monday	Tuesday	Wednesday	Thursday	Friday
TIME	Paperwork	Student Interactions		Model and Practice	Model and Practice
9:00 AM	Intro to Noyce (*tH* Master Teacher)	Pre and Post Camp Prep *(All)*	Arduinos PD *(tH Alumni/Mentor)*	Week 1 Lesson Plans *(tH Master Teacher)*	Week 2 Lesson Plans *(tH Master Teacher)*
9:30 AM	Counselor for Campers *(tH Mentors)*				
10:00 AM					
10:30 AM	Day in the life of a counselor (tH Master Teacher)				
11:00 AM	Camp Theme Overview and Camp Schedule (tH Master Teacher)		Cultural Competence *(tH Master Teacher)*		
11:30 AM					
12:00 PM	Lunch	Lunch	Lunch	Catered Lunch	Catered Lunch
12:30 PM					
1:00 PM	Icebreaker/Team-Building *(tH Mentors)*	Icebreaker/Team-Building *(tH Mentors)*	Opening Day Skit *(Team Leads)*	Opening Day Skit *(Team Leads)*	Opening Day Skit *(Team Leads)*
1:30 PM	Group Norms *(Team Leads)*	Opening Day Skit *(Team Leads)*			
2:00PM					Practice Teaching Week 2 Lesson

Source: Noyce Internship Institute Training Guide for Week One of Professional Development.

Table 3. NOYCE Intern Week Two Training Overview.

NOYCE Intern Week Two Training Overview				Sponsored by: *teach* HOUSTON
	Monday	Tuesday	Wednesday	Thursday
Time	Technology part 2			
9:00 AM	New Teachers PD *(tH Alumni/Mentor)*	Culturally Responsive Pedagogy (CRP) *(tH Alumni/Mentor)* Practice Teach *(tH Interns)*	Camp Preparation *(All)*	Camp Preparation *(All)*
9:30 AM				
10:00 AM	Break			
10:30 AM	Tech Integration PD *(tH Aulumni/Mentor)*			
11:00 AM				
11:30 AM	Flex time (pre/post prep; materials prep; practice)			
12:00 PM	Lunch	Catered Lunch	Lunch	Lunch
12:30 PM				
1:00 PM	Maker Space	Maker Space	"STEMifying Life Science" PD *(tH Alumni/Mentor)*	Meet the Campers *(tH Alumni/Mentor)*
1:30 PM				
2:00PM			Final Skit Rehearsal *(Team Leads & Interns)*	Camp Orientation *(All)*
2:30PM				
3:00PM				
3:30 PM			Team Building Activity *(Team Leads & Interns)*	Final Prep *(Team Leads & Interns)*
4:00 PM				

Source: Noyce Internship Institute Training Guide for Week Two of Professional Development.

For example, the theme of CRP was incorporated into many of the Internship Institute sessions such as *Cultural Competence, Culturally Responsive Pedagogy, Building Relationships, and Team Building Activities*. Students start their journey by engaging in a "What is Culture" activity. This is where students began the reflexive process of understanding the various aspects of culture and how culture informs instruction. In the "Funds of Knowledge" activity, students explored more deeply their own funds of knowledge and how culture can be used as a cognitive scaffold. In the "Tale of Two Nora's" Activity, students analyzed lesson plans for CRP and also distinguished between two different styles of classrooms, one inclusive of CRP and one not inclusive of CRP. Finally, students engage in a "Chalk Talk" activity where they delve deeper into CRP and compare their thoughts with others.

INTERNSHIP: A PATHWAY TO THE TEACHING PROFESSION

Some researchers have found that internships are not effective for recruiting students into STEM teacher education programs (Worsham, Friedrichsen, Soucie, Barnett, & Akiba, 2014) while others have shown that they are useful for offering additional career preparation (Tomanek & Cummings, 2000). Internships can offer the positive engagement that is important for recruitment and retention of students to the teaching education program and the teaching profession. Formal and informal activities which are part of internships are critical components of teacher education programs as they help preservice teachers develop and build their teaching and professional skills which can improve their self-efficacy (Tuchman & Issacs, 2011) and establish their professional identity (Korthagen, 2004).

Internships offer experiences which engage preservice teachers in activities which are significant for forming commitments to the teaching profession. Many of the activities are similar or identical to those the preservice teachers will be engaging in with their future classrooms. Exposure through internships is beneficial in that it will allow preservice teachers to evaluate and assess what it will take to be an effective STEM teacher for their students. This is true of the Noyce Scholarship Program, where participants' testimonies indicate its importance, including the Noyce Internship Program to their teacher training. For example, Student 5 stated:

> All Noyce-related events represent a fantastic opportunity that allows me to accomplish the following: (1) learn something new that can have a significant impact in my students; (2) to energize me about teaching; and (3) to network with colleagues.

Additional references to the impact of the Noyce Internship Program are discussed in Chapter 3, which outlines the narratives of three program participants.

Evaluation and assessment of the Noyce Internship Program indicated that its components, professional development sessions, and STEM camp facilitated and guided the preservice teacher on the pathway to the teaching profession early in their academic career (Zhao & Zhang, 2017). The professional development helped the Noyce interns make a direct connection to activities they would be involved in as they moved into the teaching profession in the 7–12 grade level. It allowed for the formation of a learning community for which the interns could rely on once in the field. By serving as facilitators in the STEM camp, interns put theory into practice as they developed into teachers, innovators, and mentors. The Noyce Internship Program also offered positive experiences through teacher/learning with lesson and curriculum development and hands-on STEM teaching activities to 5th–9th grade students as they engaged in authentic, real-world student teaching. The internship informed the preservice teachers of what to expect in the classroom as a teacher.

CONCLUSION

Informal and formal experiences in a teacher education program are important for building teacher identity and confidence. Preparation beyond the traditional course content is needed to train effective STEM teachers who will thrive in the teaching profession. *teach*HOUSTON embeds additional training opportunities within its programming to ensure that the preservice teachers develop the skills required to make them well-rounded teachers. Engaging in formal and informal activities of courses, professional development, lab-based research, and internships, preservice can set the stage for preservice teachers' teaching philosophy and persistence in the field (Korthagen, 2004; Mockler, 2011). Participation in these activities early on and throughout enrollment in the teacher preparation program will guide preservice teachers and help them realize what/how things occur in the classroom in terms of the learning as well as social environment, and how to deal with various situations. This is particularly important for training STEM teachers as research shows that those interested in being a STEM teacher have had positive experiences in the classroom as a student with a science teacher and/or working with K–12 students (Westerlund, Radcliffe, Smith, Lemke, & West, 2011). Positive experiences must continue throughout the teacher preparation program in their learning environment, i.e., in their courses and in experiences with K–12 students. For the *teach*HOUSTON program, overall, the impact of exposure to informal and formal activities associated with internships as well as the other various training opportunities has led to improved teacher self-efficacy, content knowledge, and motivation (Stokes et al., 2017). The importance/impact of the Noyce Internship Program on teacher training was informed by literature and verified through the testimonies of interns. The Noyce Internship Program model has proven to be a successful method of training preservice teachers and may be used by other teacher education programs for preparing effective STEM teachers.

AUTHORS' NOTE

This research was funded by the following grants:

- National Science Foundation DUE Award 1240083, *Recruitment, Preparation, and Retention of STEM Students as High School Teachers.*
- National Science Foundation DUE Award 1557309, *University of Houston: Learning through Informal and Formal Experience.*
- National Science Foundation DUE Award 1557273, *Collaborative Research: Understanding Robert Noyce Teacher Scholarship Outcomes in Texas.*

REFERENCES

Amolins, M. W., Ezrailson, C. M., Pearce, D. A., Elliott, A. J., & Vitiello, P. F. (2015). Evaluating the effectiveness of a laboratory-based professional development program for science educators. *Advances in Physiology Education, 39*(4), 341–351.

Aslam, F., Adefila, A., & Bagiya, Y. (2018). STEM outreach activities: An approach to teachers' professional development. *Journal of Education for Teaching*, *44*(1), 58–70.

Baker, W., & Keller, J. (2010). Science Teacher and Researcher (STAR) Program: Strengthening STEM education through authentic research experiences for preservice and early career teachers. *Peer Review*, *12*(2), 22–26.

Ball, D. L., & Cohen, D. K. (1996). Reform by the Book: What is—or might be—the role of curriculum materials in teacher learning and instructional reform? *Educational Researcher*, *25*(9), 6–14. doi:10.3102/0013189X025009006

Board of Teacher Registration, Queensland. (2003). *The multiple faces of internships: Report of phase one of a BTR project on internships in preservice teacher education*. Toowong, QLD: Board of Teacher Registration.

Brown, S. L., Bolton, K., Chadwell, N., & Melear, C. T. (2002). Preservice secondary science teacher apprenticeship experience with scientists. In *Proceedings of the annual international conference of the association for the education of teachers in science*. Charlotte, NC: AETS.

Coenders, F., Terlouw, C., Dijkstra, S., Pieters, J., & Pieter, J. (2010). The effects of the design and development of a chemistry curriculum reform on teachers' professional growth: A case study. *Journal of Science Teacher Education*, *21*(5), 535–557. doi:10.1007/s10972-010-9194-z

Colby, S. L., & Ortman, J. M. (2014). *Projections of the size and composition of the US population: 2014 to 2060, current population reports*. P25-1143. US Census Bureau, Washington, DC.

College Board. (2008). *Coming to our senses: Education and the American future. Report of the commission on access, admissions and success in higher education*. College Board, New York, NY. Retrieved from http://professionals.collegeboard.com/profdownload/coming-to-our-senses-college-board-2008.pdf

Council for the Advancement of Standards in Higher Education. (2006). *Internship programs*. Washington, DC: Naylor Association Solutions.

Darling-Hammond, L. (2006). *Powerful teacher education: Lessons from exemplary programs* San Francisco, CA: Jossey-Bass.

Dean, L. A. (Ed.). (2006). *CAS professional standards for higher education*. Washington, DC: Council for the Advancement of Standards in Higher Education.

Du, W., Liu, D., Johnson, C. C., Sondergeld, T. A., Bolshakova, V. L. J., & Moore, T. J. (2019). The impact of integrated STEM professional development on teacher quality. *School Science & Mathematics*, *119*(2), 105–114. doi:10.1111/ssm.12318

Gardner, D. (1983). *A Nation at Risk: The imperative for educational reform. An open letter to the American people. A report to the nation and the secretary of education*. Washington, DC: Department of Education. National Commission on Excellence in Education.

Gay, G. (2010). *Culturally responsive pedagogy: Theory, research, and practice*. New York, NY: Teachers College Press.

Hord, S. M. (1997). *Professional learning communities: Communities of continuous inquiry and improvement*. Austin, TX: Southwest Educational Development Laboratory.

Hunzicker, J., Biddison, T., Hickey, M., & McClethen, A. (2014). Four reasons why pre-service and practicing teachers should present at professional conferences. *MLET: The Journal of Middle Level Education in Texas*, *1*(1), 4.

Jackson, C. D., & Mohr-Schroeder, M. J. (2018). Increasing STEM literacy via an informal learning environment. *Journal of STEM Teacher Education*, *53*(1), 41–52.

Kim, M. S., & Keyhani, N. (2019). Understanding STEM teacher learning in an informal setting: A case study of a novice STEM teacher. *Research and Practice in Technology Enhanced Learning*, *14*(1), 9. doi:10.1186/s41039-019-0103-6

Korthagen, F. A. J. (2004). In search of the essence of a good teacher: Towards a more holistic approach in teacher education. *Teaching and Teacher Education*, *20*(1), 77–97. doi:10.1016/j.tate.2003.10.002

McCarthy, R. (2011). Stewart & Ray's big adventure: A research experience for teachers at UMASS/Amherst. *Technology and Engineering Teacher*, *71*(3), 22–26.

McDermott, L., Shaffer, P., & Constantinou, C. (2000, November 1). Preparing teachers to teach physics and physical science by inquiry. *Physics Education*, *35*(6), 411–416.

McFadden, J. R., & Roehrig, G. H. (2017). Exploring teacher design team endeavors while creating an elementary-focused STEM-integrated curriculum. *International Journal of STEM Education, 4*(1), 1–22. doi:10.1186/s40594-017-0084-1

Melear, C. T., Goodlaxson, J. D., Warne, T. R., & Hickok, L. G. (2000). Teaching preservice science teachers how to do science: Responses to the research experience. *Journal of Science Teacher Education, 11*(1), 77–90.

Mockler, N. (2011). Beyond: 'what works': Understanding teacher identity as a practical and political tool. *Teachers and Teaching, 17*(5), 517–528. doi: 10.1080/13540602.2011.602059

National Association of Colleges and Employers. (2011). Position statement on U.S. Internships: A definition and criteria to assess opportunities and determine the implications for compensation. Retrieved from https://www.naceweb.org/connections/advocacy/internship_position_paper/

National Research Council. (1996). *National science education standards*. Washington, DC: The National Academies Press. doi:10.17226/4962

O'Neill, N. (2010). Internships as a high-impact practice: Some reflections on quality. *Peer Review: Emerging Trends and Key Debates in Undergraduate Education, 12*, 4–8.

Pop, M. M., Dixon, P., & Grove, C. M. (2010). Research experiences for teachers (RET): Motivation, expectations, and changes to teaching practices due to professional program involvement. *Journal of Science Teacher Education, 21*(2), 127–147.

Pozo, R., Martinez-Aznar, M., Rodrigo, M., & Varela, P. (2004). A comparative study of the professional and curricular conceptions of the secondary education science teacher in Spain: Possible implications for ongoing teacher education. *European Journal of Teacher Education, 27*(2), 193–213.

Russell, S. H., & Hancock, M. P. (2007a). *Summary: Evaluation of the research experiences for teachers (RET) program: 2001–2006*. Menlo Park, CA: SRI.

Russell, S. H., & Hancock, M. P. (2007b). *Final report: Evaluation of the research experiences for teachers (RET) program: 2001–2006*. Menlo Park, CA: SRI.

Southerland, S. A., Granger, E. M., Hughes, R., Enderle, P., Ke, F., Roseler, K., & Saka, Y. (2016). Essential aspects of science teacher professional development: Making research participation instructionally effective. *AERA Open, 2*(4), 1–16. doi:10.1177/2332858416674200

Stokes, D., Evans, P., & Craig, C. (2017). Developing STEM teachers through both informal and formal learning experiences. In J. Mena, F. J. García-Peñalvo, A. García-Valcárcel, & M. M. Del Pozo (Eds.), *Search and research: Teacher education for contemporary context*. Salamanca: Ediciones Universidad de Salamanca.

Terrazas-Marín, R. A. (2018). Developing non-formal education competences as a complement of formal education for STEM lecturers. *Journal of Education for Teaching, 44*(1), 118–123. doi: 10.1080/02607476.2018.1422613

Tomanek, D., & Cummings, K. E. (2000). The use of secondary science classroom teaching assistant experiences to recruit academically talented science majors into teaching. *Science Education, 84*, 212–227.

Tuchman, E., & Issacs, J. (2011). The influence of formal and informal formative pre-service experiences on teacher self-efficacy. *Educational Psychology: An International Journal of Experimental Educational Psychology, 31*(4), 413–433.

Waight, N., & Abd-El-Khalick, F. (2012). Nature of technology: Implications for design, development, and enactment of technological tools in school science classrooms. *International Journal of Science Education, 34*(18), 2875–2905. doi:10.1080/09500693.2012.698763

Wenglinsky, H., & Silverstein, S. C. (2007). The science training teachers need. *Educational Leadership, 64*, 24–29.

Westerlund, J. F., Radcliffe, R. A., Smith, D. A., Lemke, M. R., & West, S. S. (2011). Profiles of US science teacher candidates: Safeguards amidst the gathering storm. *International Journal of Environmental & Science Education, 6*(3), 213–227.

Westerlund, J. F., Schwartz, R. S., Lederman, N. G., & Koke, J. R. (2001). *Teachers learning about nature of science in authentic science contexts: Models of inquiry and reflection*. Annual Meeting of the National Association for Research in Science Teaching, St. Louis, MO.

Wilhelm, J., & Fisher, M. H. (2019). Creating academic teacher scholars in STEM education by preparing preservice teachers as researchers. In B. Doig, J. Williams, D. Swanson, R. Borromeo Ferri, & P. Drake (Eds.), *Interdisciplinary mathematics education* (pp. 281–296). Cham: Springer.

Worsham, H. M., Friedrichsen, P., Soucie, M., Barnett, E., & Akiba, M. (2014). Recruiting science majors into secondary science teaching: Paid internships in informal science settings. *Journal of Science Teacher Education*, *25*(1), 53–77. doi:10.1007/s10972-013-9360-1

Zhao, H., & Zhang, X. (2017). The influence of field teaching practice on pre-service teachers' professional identity: A mixed methods study. *Frontiers in Psychology*, *8*, 1264. doi:10.3389/fpsyg.2017.01264

Zientek, L. R., & Thompson, B. (2008). Preparing high quality mathematics and science teachers: Are we meeting the challenge? *Research in the Schools*, *15*, 1–19.

EXAMINING THE IMPACT OF INFORMAL EXPERIENCES ON PRESERVICE TEACHERS' SELF-EFFICACY

Paige K. Evans, Leah McAlister-Shields, Mariam Manuel, Donna W. Stokes, Ha Nguyen and Cheryl J. Craig

ABSTRACT

This chapter illuminates the impact of providing informal learning experiences for students pursuing Science, Technology, Engineering, and Mathematics (STEM) teaching careers at a time when there is a considerable shortage of qualified teachers in America's urban centers. Preservice STEM teachers were provided with the opportunity to participate in a National Science Foundation (NSF) grant funded Noyce Internship Program prior to serving as counselors and teaching assistants in a STEM camp for underrepresented middle school students. Through the Noyce Internship Institute, participants were introduced to interactive sessions that model promising teaching practices including inquiry-based and project-based learning. This narrative inquiry examines the impact of these experiences on preservice STEM teachers' self-efficacy and highlights outcomes in three areas: increase of preservice teachers' confidence, classroom management, and strengthening their desire to teach STEM.

Keywords: STEM education; informal learning; narrative inquiry; preservice teachers; self-efficacy; teacher self-efficacy; preservice teacher program; STEM internship; *teach*HOUSTON

Preparing Teachers to Teach the STEM Disciplines in America's Urban Schools
Advances in Research on Teaching, Volume 35, 85–108

ISSN: 1479-3687/doi:10.1108/S1479-368720210000035006

Science, technology, engineering, and mathematics (STEM) education is essential for the United States to remain globally competitive in the twenty-first century, which, in turn, requires special investments in today's youth (National Academy of Sciences, 2007). Interest in STEM content should be encouraged at a young age and must be sustained throughout their K–12 schooling experiences. Moreover, a sound understanding of STEM concepts strengthens the skills that people use every day such as solving problems creatively, thinking critically, working cooperatively, and using technology effectively. Unfortunately, progressive instructional practices are still uncommon in American science classrooms, despite efforts to encourage reform in science instruction (National Research Council, 2005; Windschitl, 2009). Accordingly, investment of resources in teacher development is essential to help current and future educators embrace student-centered pedagogical methods. Teachers that promote inquiry, creativity, and collaboration in the classroom ultimately help prepare students to become productive members of society (Darling-Hammond, 2012).

Novice math teachers claim that their preservice teacher preparation programs do not prepare them to overcome feelings of inadequacy when teaching in high-poverty school districts (Zientek & Thompson, 2008). Hence, for teachers to be impactful in high need schools, it is imperative that programs provide ongoing formal and informal experiences that create opportunities for aspiring teachers to build skills and heighten their self-efficacy as educators (Tuchman & Isaacs, 2011). Informal learning is described as learning that takes place outside of the bounds of a classroom where students traditionally are instructed (Sullenger, 2006). Moreover, allowing preservice teachers to participate in these environments empowers them to apply their content learning outside of a university classroom setting (Jackson, Mohr-Schroeder, & Little, 2014).

COMING TO THE QUESTION

Teachers' sense of self-efficacy can be heightened by exposure to multiple forms of formal and informal teaching experiences. Doing so is critical because a strong sense of efficacy has been correlated with improved student achievement outcomes and teacher retention (Anderson, Green, & Loewen, 1988; Midgley, Feldlaufer, & Eccles, 1989, pp. 981–992; Moore & Esselman, 1992). Informal learning environments have been shown to positively affect the confidence levels of preservice teachers and their attitudes toward STEM-related subject areas (Jackson et al., 2014; Stokes, Evans, & Craig, 2017). Still, university-based preservice teacher preparation programs continue to offer limited opportunities for growth through experience in informal settings. In this study, preservice teacher participants were offered grant-funded internships which provided both formal and informal learning opportunities. Initially, preservice STEM teachers participated in a Noyce Internship Institute where they were introduced to interactive sessions that modeled promising teaching practices including inquiry-based and project-based learning. Subsequently, participants served as counselors and teaching assistants in a STEM summer camp for middle school students predominantly from underrepresented populations.

The purpose of this narrative inquiry is to examine the impact of informal learning experiences (ILEs) on preservice STEM teachers' self-efficacy along with their confidence, motivation, and commitment to the teaching profession, and classroom management at the University of Houston, a large urban minority serving institution in Texas. For the purpose of this study, classroom management was defined as the skills and strategies an instructor implements in order to cultivate a classroom environment that helps ensure student success. The results of this inquiry will inform our personal practice as leaders in preparing undergraduate STEM teachers in diverse, urban settings and contribute to the science teacher education and the general teacher knowledge base nationally and internationally. We begin our study with our theoretical framework followed by our research methodology. Additionally, we will present the stories of preservice teachers who engaged in the internship institute and then worked with underrepresented middle school students in a STEM summer camp. Finally, we conclude with a discussion of implications and future steps.

RESEARCH QUESTION

The research question that guides this study is *what is the impact of informal experiences on preservice teachers' self-efficacy*?

THEORETICAL FOUNDATION

We now unearth the theoretical underpinnings pertaining to teacher self-efficacy as it stems from social cognitive theory.

Social Cognitive Theory

The theoretical framework for this study is grounded in Bandura's Social Cognitive Theory (SCT) (Bandura, 1997). SCT posits a model for learning that involves reciprocal interactions between personal, behavioral, and environmental determinants (Pajares, 1996). SCT views individuals as active contributors in their own development with agency to derive change in behaviors (Bandura, 1986). The construct of self-efficacy was first introduced by Bandura in 1977 and defined as the "beliefs in one's capabilities to organize and execute the courses of action required to produce given attainments" (Bandura, 1977, p. 3). The formation of this construct was driven by Bandura's assertion that human functionality extends beyond knowledge and skills acquired over time (Siwatu, 2007). Bandura (1986) theorized that action is generated by an individual's belief in their ability to produce desired outcomes.

Self-efficacy stems from the SCT established by Bandura (1997) who defined it as "people's beliefs about their capabilities to produce designated levels of performance that exercise influence over events that affect their lives" (Bandura, 1994, p. 2). Self-efficacy serves as a regulator of individual and group behavior as it manages personal and environmental factors. Groups and individuals utilize their efficacy beliefs "as regulatory mechanisms to engage cognitive, motivational, affect, and selected processes that translate competence into successful action behavior or performance" (Rew, 2013, p. 32). These processes work in conjunction and serve as mediators between an individual and group's efficacy beliefs and displayed behavior (Bandura, 1995, 1997, 1999).

Self-efficacy is the driving force behind the way people perceive themselves, think, behave, and find motivation (Bandura, 1997). A strong sense of efficacy influences human achievements through various aspects including education, health, sports, and work (Bandura, 1997). Bandura (2001) also suggests that in addition to performance, an individual's level of persistence and effort are also correlated to teacher efficacy beliefs. Teacher efficacy is defined as a teacher's belief that his or her actions can impact student learning (Guskey & Passaro, 1994). A considerable amount of research supports the direct connection between increased teacher self-efficacy and improved student achievement (Skaalvik & Skaalvik, 2007; Tschannen-Moran & Hoy, 2001; Wolters & Daugherty, 2007). Additionally, high levels of teacher efficacy can yield greater teaching effort and persistence through challenging situations (Tschannen-Moran & Hoy, 2001). Positive efficacy beliefs also allow teachers to employ innovative instructional techniques (Gibson & Dembo, 1984). Hence, SCT and the construct of self-efficacy have served as a prominent framework to guide research on teacher beliefs and how they influence instructional practices. Furthermore, scholars have argued that providing preservice teachers with both formal and ILEs can significantly heighten their efficacy beliefs (Tuchman & Isaacs, 2011). Thus, it is paramount to acquire a deeper understanding of how ILEs influence and shape aspiring teachers' efficacy beliefs. As such, this narrative inquiry investigates the perceptions of confidence and competence reported by preservice teachers upon participation in informal teaching and mentoring experiences with middle school aged students through a STEM summer camp.

REVIEW OF LITERATURE

Defining Informal Learning Experiences

The term "informal learning" is referred to as the type of learning that takes place outside the formalized boundaries of classroom settings (Dierking, Falk, Rennie, Anderson, & Ellenbogen, 2003). Different to formal learning, in an informal environment, learning is presented in a more engaging and hands-on way (Roberts et al., 2018). ILEs enable preservice teachers to connect the theoretical context covered in coursework to a learning environment that is not a classroom.

Through these opportunities, preservice teachers are able to implement the instructional methods modeled and discussed during their programmatic activities (Harlow, 2012). Researchers contend that teaching science through informal contexts may result positively on teaching candidates' "enjoyment of science and understanding of how inquiry-based strategies develop conceptual knowledge" (Wallace & Brooks, 2014, p. 178). Moreover, informal learning also allows prospective teachers to attain a deeper understanding of children's ideas about the content, and, thus, they gain insight into students' thinking about the world around them (Harlow, 2012). Participation in such experiences provides opportunities for preservice teachers to apply the theoretical knowledge gained by their formal learning experiences with hands-on teaching and mentoring experiences.

Informal Learning Experiences in STEM

ILE in STEM occurs in an environment that facilitates and promotes STEM education in a nontraditional classroom setting. Kotys-Schwartz, Besterfield-Sacre, and Shuman (2011) classify three major settings for informal learning: (1) everyday experiences; (2) designed settings which occurs in places such as museums, zoos, aquariums, and environmental centers; and (3) programmed settings, which take place in various venues including K–12 schools, universities, and other informal establishments. Dierking et al. (2003) notes that learning in science rarely develops from a single experience, rather it emerges overtime, derives from real-world experience and within a diverse physical and social context. These dynamic situations and interactions, which are the core of informal learning, influence the ways individuals construct scientific knowledge, attitudes, behaviors, and understanding (Dierking et al., 2003).

Research has shown that ILEs in STEM are beneficial to students. In a quantitative study using a two-way analysis of variance (ANOVA) statistical analysis, researchers concluded that informal learning environments improve students' scientific reasoning abilities (Gerber, Cavallo, & Marek, 2001). Opportunities for informal learning advance students' STEM content acquisition and promote teamwork, critical thinking, and leadership skills (Martin, 2004; Newell, Zientek, Tharp, Vogt, & Moreno, 2015; Yilmaz, Ren, Custer, & Coleman, 2010). Informal STEM learning experiences inspire students to learn, retain, and pursue careers in STEM fields (Denson, Austin, Hailey, & Householder, 2015; Mohr-Schroeder et al., 2014; Roberts et al., 2018). Given the nation's shortage of STEM workforce and the lack of diversity in the STEM fields, ILEs might provide a new avenue for learning about STEM topics and exposure toward STEM careers for students, especially those from underrepresented populations (Denson et al., 2015).

Informal Learning Experiences in Preservice Teacher Preparation

Researchers have asserted that if we are to expect our future teachers to think and teach in innovative ways then we must allow them to experience learning in different settings (Putnam & Borko, 2000). Still, the majority of professional development regarding ILEs is aimed toward in-service teachers, thus, leaving a gap in the preparation of prospective teachers (Phillips, Finkelstein, & Wever-Frerichs, 2007). Consequently, researchers and educators have strongly voiced their support for university-based teacher preparation programs to include ILEs (i.e., Anderson, Lawson, & Mayer-Smith, 2006; DeWitt & Storksdieck, 2008). Professional organizations have also taken a stance on this issue and expressed the need for the inclusion of informal learning. The American Educational Research Association (AERA) formed an Informal Learning Environments

Research Special Interest Group, and the National Science Teachers Association (NSTA) released a position statement supporting the role of informal learning in science education (Olson, Cox-Petersen, & McComas, 2001). Furthermore, the National Association of Research in Science Teaching (NARST) developed an Informal Science Education Ad Hoc Committee charged with describing, supporting, and researching science education in informal settings.

Science Education

Harlow (2012) studied the impact of integrating informal science learning experiences through community family science nights into a preservice teacher education program. The study involved observations of 46 preservice teachers as they prepared, participated, and reflected upon their experiences for a family science night. One of the focus areas for research was the impact it had on preservice teachers' confidence to teach science. The findings supported the author's assertion that engaging students in meaningful, informal, educational activities can serve as enjoyable experiences for preservice teachers and excite them about the prospect of teaching science content. With regards to heightening their confidence in their ability to teach science, Harlow stated:

> Whether this one event led to greater self-confidence in the preservice teachers' ability to teach science to a classroom of students is not clear from the data. Nevertheless, based on this one successful event, the preservice teachers may begin to see themselves as people who can engage children in successful and enjoyable science. (p. 217)

Harlow further spoke of the value of informal learning by discussing the implications it has on how science can be learned. During these experiences, the students and adults have a greater sense of agency versus the traditional classroom environments, thus, allowing the activities to become more self-directed and student-driven. This argument is also supported by Putnam and Borko (2000) who claimed that the classroom setting can be constraining and as such, "engaging in learning experiences away from this setting may be necessary to help teachers 'break set' to experience things in new ways" (p. 6).

Mathematics Education

According to Popovic and Lederman (2015) research regarding the use of ILEs in mathematics education is in its infancy, this applies to in-service professional development and teacher preparation programs. However, the authors discussed the recent efforts that have been made to provide preservice teachers with opportunities to experience mathematics in settings outside of a formal classroom in an effort to, "enhance their understanding of applications of their mathematics curriculum" (p. 130). One such example includes a study conducted by Johnson and Chandler (2009) which found that preservice mathematics teachers were able to draw connections between their mathematics coursework and real-world applications. Specifically, the participants enrolled in a methods course were asked to visit the USS North Carolina Battleship in North Carolina on Cape Fear Coast. The prospective teachers were then charged with developing a lesson plan that provided instruction through the perspective of this Battleship trip. The authors explained that the participants were not only successful in building connections but also expressed the value they perceived in providing students with opportunities to visit informal sites as a teaching tool to promote discovery and application to real-world scenarios.

This notion is further supported by Popovic and Lederman (2015) who utilized an ethnographic approach to investigate the depth of mathematical concepts that preservice and in-service mathematics teachers recognized in informal sites with respect to the Common Core Standard for Mathematical Practice 4: Model with Mathematics. The researchers found that although initially the teachers were able to recognize some concepts at the surface level, such as numbers, shapes, or figures, it was not until after the informal learning took place that they were also able to make connections with the more abstract concepts in mathematics. Thus, the authors recommend use of ILEs in order to create opportunities for teachers "to identify mathematics represented in the real world as well as to engage in mathematical modeling of real-world phenomena" (p. 139).

RESEARCH METHOD

Grounded in the method of narrative inquiry (Clandinin & Connelly, 2000; Craig, 2011; Polkinghorne, 1995), this research study is an inquiry into the experiential narratives of preservice STEM teachers who attend the University of Houston, a tier-one research institution in the fourth largest urban city in America. Their lived stories convey the experiences encountered in the Noyce Internship Program which includes the Noyce Internship Institute and an internship in the summer STEM camp for underrepresented middle school students. These narratives form the primary data source for this study.

Narrative inquiry is strongly influenced by John Dewey (1938) who believed that one must rely on past experiences and knowledge to solve problems. Dewey believed that life is experience and that experience is, in fact, education. Further to this, Clandinin and Connelly (2000) maintained that "Experience happens narratively. Narrative inquiry is a form of narrative experience. Therefore, educational experience should be studied narratively" (p. 19). The process of narrative inquiry involves "a collaboration between the researcher and the participants, over time, in a place or series of places, and in social interactions and with milieus" (p. 20). We relied on preservice science teachers to tell their stories as STEM students participating in the NSF-funded informal learning program in an attempt to understand their experiences.

Narrative inquiry occurs within a "three-dimensional narrative inquiry space" which allows us to travel in the following directions: "inward, outward, backward, forward, and situated within the place" (Clandinin & Connelly, 2000, p. 49). This inquiry took us back as former students, and now as contributors to the field in reflecting on our personal experiences of learning in STEM through inquiry. Hence, the narrative inquiry was integrated in the study to contribute to the data collection in STEM secondary education. By illustrating multiple identities in

STEM, we gather the values, beliefs, and attitudes that may contribute to the overlooked or unnoticed experiences in STEM. Our intention is to share the distinct stories found that can influence research, practice, or theory in the future.

Sources of Evidence

The central data-yielding approach for this study was gathered qualitatively through multiple rounds of focus group interviews. Qualitative research provides an array of different perspectives that can be complex in nature. It focuses on exploring and provides a richness of details to understand the experiences and beliefs of participants.

Students from the *teach*HOUSTON program, the University of Houston's secondary STEM teacher preparation program, were asked to participate in the research study. Participants were purposefully recruited via emails and a website. The research team conducted two rounds of focus group interviews. In the first round, a pilot study including a one-hour focus group interview with three Noyce interns was conducted by faculty members and two graduate assistants. To capture the full meaning, the interview was recorded, transcribed, and analyzed by faculty members. The themes that emerged from the pilot interview informed the revision of the interview protocol for the second round of group interviews. In the second round, two graduate assistants conducted the interviews with three *teach*HOUSTON preservice teachers who completed the Noyce Internship program. The second focus interview was also recorded, transcribed, and analyzed using the Dedoose qualitative software program. Data relating to participants' discussion of self-efficacy and the impact of informal learning on preservice teacher self-efficacy were identified. Researchers created categories and looked across cases to identify patterns or emerging themes (Craig, Evans, Verma, Stokes, & Li, 2019; Craig et al., 2018).

Interpretive Tools

Interpretive tools were employed to support the narrative inquiry including broadening, burrowing, and storying and restorying (Connelly & Clandinin, 1990). We expanded beyond the immediate study by broadening which allowed for a comprehensive representation of the data; in order to discern larger meanings brought to the study, we employed burrowing which allowed us to look intensively at the participant's experience; we were able to show how change happens at different levels, both micro and macro, through storying and restorying. As Connelly and Clandinin explain,

> As we engage in a reflective research process, our stories are often restoried and changed as we, as teachers and/or researchers 'give back' to each other's way of seeing our stories (1990, p. 9).

SITUATING THE INQUIRY

The teachHOUSTON Program

*teach*HOUSTON is a secondary educator preparation program that was created in 2007 to develop highly qualified mathematics and science teachers. Students from 11 different majors in the College of Natural Science and Mathematics participate in the *teach*HOUSTON program. In addition to a major in mathematics or an area of science, students complete 23 hours of education coursework to include a semester of apprentice teaching. Education courses in the *teach*HOUSTON program consist of four field-base classes where students develop and facilitate math and science lessons in secondary classrooms and three theory-based courses to inform and support preservice teacher instruction. In addition to the formal learning experienced by preservice teachers in the *teach*HOUSTON program, a number of ILEs are also provided which include summer camp counseling, conference attendance, and internships.

The Noyce Internship Institute

The grant-funded Noyce Internship Institute provides professional development to preservice teachers as they prepare to become teaching assistants in a Summer STEM Camp for underserved middle school students. This allows preservice teachers to engage in meaningful ILEs which further builds their capacity as future educators and their desire to impact change in the lives of students. The professional development modules included in the institute provide interactive training sessions that model promising teaching practices and classroom management to increase student engagement and success. Topics included the following: Working with Middle School Students; Professionalism; What is Facilitation?; Classroom Management; Engineering Design; Inquiry-based Learning; College and Career Readiness; Bullying; Technology; Brain Building Lesson Planning; and Growth and Fixed Mindset. Faculty members from the University led the sessions as well as lead interns who had previous experience as Noyce Interns and showed a capacity for leadership. These sessions also built camaraderie with fellow preservice teachers as they included icebreakers and sessions specifically developed for team building.

The Summer STEM Camp

The summer STEM camp is a 2-week academic program, which takes a committed role in shaping the education of underserved, middle school students by cultivating their skills in STEM education on a college campus. The core curriculum for the camp incorporates engineering design-based projects that support innovation, prototyping, and twenty-first century skills. The curriculum consists of an integrated approach to STEM topics that include the following: Technology, Communications, Field Excursions, Life Sciences, Engineering, and Mathematics. Consequently, Noyce Interns engage with middle school students in activities and projects that promote critical thinking, problem-solving, and collaboration.

NARRATIVES OF EXPERIENCE

After the ILE, three participants who were part of the summer internship program were interviewed. To assure confidentiality, pseudonyms were used instead of the participants' birth names. Bernardo, Raul, and Ana were part of the NSF Noyce grant program. What follows are the experiences of each of our three participants.

Bernardo's Account

Bernardo identifies as a male. He is a sophomore and majors in mathematics. From the beginning of his college career, he knew that he wanted to become a teacher. Subsequently, he found the *teach*HOUSTON program and realized that this was his home. When asked what he was planning to teach, he excitedly shared that "I plan to teach high school math." Our discussion about Bernardo's ILE started with his description of a teacher that made a difference in student lives. Pausing for a moment, Bernardo expressed his appreciation for his high school math teacher who cared a lot about all the students: "*Mr. Dot taught my AP Calculus… He was always putting student needs before his needs.*" Bernardo observed that while many other teachers were just "*Okay, I'm just going to get our stuff out of the way and just like, get all this information [sped] up,*" his high school math teacher would "*hold things off and wait until he was sure that everyone in the class understood… If [he] noticed that someone was feeling down or anything, he would ask – are you okay?*"

As our conversation continued, Bernardo described in depth his experience in the summer Noyce Internship program. At the start of the internship, Bernardo did not feel that he made a difference in students' lives because he felt that middle school students have different motivators:

> In elementary schools, kids are mostly trying to please the adults in their lives, where high school kids are trying to get into college, [but] in middle school students like lack both of those. So, I felt like they were just like, oh I'm here because my mom made me come here.

Because of these interactions with middle school students during the program, Bernardo decided that he wanted to make a lasting impact in the lives of students. He said that by the end of the internship, he was able to make connections with students "*a lot easier.*" He proudly declared "*even though teaching middle school is not my first choice, I feel if I do land in that position, [it is] something I could definitely do and manage.*"

As he spent more time in the internship, Bernardo realized that these valuable experiences have changed his perception of the school and community where he was teaching. He recalled at the beginning of joining *teach*HOUSTON, his mom painted a negative picture of a middle school zoned to his neighborhood. His mom made it very clear to him that he would not be going there for his field experience. "*I was so scared. Because my mom… She made it sound so bad… She made it seem like this is really like a bad place.*" So, he said prior to serving as an intern he was very much against the idea of working with middle school students in that particular area. However, as Bernardo advanced through the program, to his surprise he was assigned to that exact middle school in his neighborhood. Yet, he shared that the experience completely contradicted what his mom told him.

> When I went into that middle school, I quickly realized that ...this is not bad at all... A lot of the kids that are there are not necessarily as labeled as the good kids. [But] they really do want to learn.

The formal field-based experience coupled with the informal internship experience helped change his perception of being a teacher, as he described:

> These kids want to learn... You just need to be like that person that pulls that desire out of them. So I've learned [that] through the internship and stuff I've learned that [there are] a lot of ways that way you can do that and that way you can motivate these kids, that way they can do their best. He happily revealed that "after this [internship], I was like, you know, this is something I could do and possibly enjoy."

As our conversation came to a close, Bernardo expressed that the experience in the informal internship program changed his perception of the teaching profession and brought value to this potential career. He explained that prior to the internship, he possessed stereotypes of teaching where the teacher would sit behind a desk and grade papers. However, as he reflected on his internship experience, he shared

> I've definitely learned [that] teaching involves a lot more work than you see... Now definitely I learned that there's so much more that goes into teaching and that it's a very valuable profession that you learn a lot from.

Raul's Account

Raul identifies as male and is a junior majoring in biology and plans to teach high school biology or chemistry once he graduates. Raul explained that he did not plan to go to college to become a teacher, rather he planned to have a career in the medical profession:

> I was a stereotypical pre-med biology student, hopeful, uh, wanting to go into medical school. But I learned that medicine isn't my thing. I actually really dislike it just because of a bad bunch of different reasons.

But when Raul needed to select a capstone for his degree, he took the *teach*HOUSTON introductory class and found it intriguing. As he progressed through the program, he recalled thinking, "*[I] really enjoyed it and that's where I made my career change.*" He explained, "*I'd rather do something that I enjoy for the rest of my life than something I really hate.*"

Our conversation started with Raul's description of teachers that are impactful in a student's life. He shared that influential teachers are those that are involved with students' lives. They play the role of a mentor who "*actually cares to hear what their students have to say.*" As he looked back at his own high school experience, he said "*there are only a few teachers that really fit that mentoring aspect.*" With this in mind, Raul said he tried his best to motivate the kids and get them excited because he was aware of the importance of being a teacher. He said "*if you're an educator in any aspect, it's not just giving them information, it's being that mentor. That's super important when you're educating or teaching someone.*"

When Raul joined the summer internship program, he was further in his *teach*HOUSTON coursework than his peers which provided him with a sense of comfort and confidence. Raul explained his approach by stating, "*I personally make it my goal to try to go around and get to know each kid individually… seeing what the kids' interests are, what they liked and what they wanted from the program.*" He was aware that it was important to "*make an effort to try to get to know the kids.*" He enthusiastically recalled an example of when he witnessed one of the students overcome challenges. According to Raul, seeing this transformation made it "*really worthwhile in the end.*" He provided additional context to the story in the explanation below:

> There was this one [student] I have in mind…I mean, he lacked confidence. That was the big thing with him. I'm like, yeah, [he has] good ideas. He just had a lack of confidence to actually put those ideas forward. Fortunately for him, he ended up winning the competition at the end. So, I feel that, you know, my mentoring [helped] in a way and I'm not going to take all of the praise for that because there was of course the internship team to help out with that. But I feel like I was a part of maybe helping him come out of his shell a little bit.

Raul's perception of the teaching profession also changed after going through the program where he learned not only pedagogical knowledge but also statistics and the state of the US educational system. Through this learning, he realized that the education system is not fully working which he explained "opens your eyes to a lot of things and tells you like hey [we] need a new batch of teachers or we need to change something." The internship also provided him a real opportunity to "constantly interact with the kids … learn how to talk to them" and to collaborate with many in-service teachers which he perceives as helpful in the transitioning period in the first two years when he begins his career. He said "the internship really did open my eyes. Having that experience is super valuable." He added that his interest changed as a result of the internship; he now actively seeks out potential opportunities for tutoring and mentoring students. He stated that he tries to

> …get a little more hands-on experience maybe over the summer or in between a spring and fall semester in January, just to get a little more experience before I am actually heading out into the field.

Ana's Account

Ana is a female junior biology major. She is of Asian descent and plans to teach high school biology or physics. Like Raul, Ana came to college with the intention of preparing for medical school. Growing up, she went to many respectable schools and naturally she felt that kids were well behaved. Ana said in her childhood, her parents "definitely had certain ideas about what teaching was and they kind of implemented that into me like growing up." When she first came in the *teach*HOUSTON program, she only wanted to see what teaching was about during the Step 1 course. Yet, like Raul, as she advanced in the *teach*HOUSTON program and the internship experience, she had more respect for what teachers do and expressed "okay, this is something that I want to do if not forever, at least for a little bit." When describing an impactful teacher, Ana said that "they are the ones that showed that they cared and ...actually wanted to hear, like if the students have any problems." She also added that they are the ones who "made sure that they were always available... they always let the students know, hey, I'm going to be here before school, after school, if you need anything during lunch, I'm here."

Ana looked back at her experience at the beginning of the internship experience and recalled feeling a lot of pressure because she was eager to make an impact in the lives of students. Even though Ana has little sisters and described being "surrounded with kids my whole life," Ana laughingly admitted that she was very stressed at the beginning of the internship institute. She explained that she was worried because at the time she had only completed the first course in the program while some of her peers were further along. Her goal was to be structured, make good lessons, and have clear learning objectives. "The making [of] lesson plans, [and] making sure everything was on time [made me] terrified."

Upon completing the internship, she felt "comfortable" as she learned how to talk to the students and to gain their trust. "That's definitely something that I feel that changed from before the internship to after." She described that the internship provided her with a real-world setting, thus enabling her to build upon her own teaching style:

> So rather than...kind of move around as much, I chose to stay with the same three kids throughout. I feel like I really got to know them a lot better and the interesting part was [that the] three of them were related and so we also had to deal with the family dynamic ... so it was kind of interesting, kind of like stepping in there and trying to take that mediator role.

Given that it all worked out, even better than she thought it would, she finds herself more motivated to work with students. She shared:

> I feel [the students'] interactions with us and the things that we talked about, like doing lunches and walks and stuff...I had one kid say I was his favorite, which was a highlight of my whole summer for the most part.

Ana emphasized that "the internship was a huge help" for her. It gave her more confidence as an aspiring teacher. For instance, she shared, "that if [in the future] any questions come up, I'll know how to direct them." She also said the experience ignited the desire to make a difference, "there's so much more that I could do" and it inspired her to be "a part of the change" for the current education system.

ANALYSIS: OVERARCHING THEMES

The purpose of this narrative inquiry is to examine the impact of ILEs on preservice STEM teachers' self-efficacy. Our findings as highlighted in Table 1 include that the internship: (1) increased preservice teachers' confidence, (2) strengthened their motivation and commitment to the teaching profession, (3) improved their classroom management, and (4) changed attitude and behavior toward teaching STEM. The findings suggest that the informal teaching

Table 1. Themes of Study and Illuminating Exemplars.

Theme	Narrative Exemplar	Participant
Increased preservice teachers' confidence	*I'd never seen a lesson plan before the internship, so that [lesson planning] really helped my confidence with creating lessons plans and during step two I was super confident with it and … now I'm like really excited.*	Ana, a female Asian preservice teacher, Interview 2019
Strengthened their motivation and commitment to teaching profession	*The first thing I did after doing the internship was look up any possible tutoring or mentoring opportunity during the summer. While I'm not in school just to gain more experience teaching outside you know the classes at [the university] because I feel like the internship really did open my eyes. [I think], having that experience is super valuable.*	Raul, a male Latino preservice teacher, Interview 2019
Improved their classroom management	*So like just kind of see like how the kids are and learn and know like okay this would be like a good strategy to use for them and this won't be.*	Bernardo, a male preservice teacher, Interview 2019
Changed attitude and behavior toward teaching in STEM	*Whatever the teacher was super passionate and interested about the subjects that they were teaching and I kind of want to convey that same type of excitement about a subject to my students.*	Ana, a female Asian preservice teacher, Interview 2019

experiences positively impact preservice teachers' self-efficacy. The teacher candidates' interviews revealed that the ILEs provided valuable benefits in their understanding of the teaching profession.

INCREASED PRESERVICE TEACHERS' CONFIDENCE

Confidence is a self-assured state of mind that has roots in self-valuing (Gauthier, 2010). Confidence is the belief in one's ability to be effective when taking on challenges (Nolan & Molla, 2017). In the teaching profession, teacher confidence is a key element of professional performance, which enables teachers to try new ideas and to be more effective in their pedagogical practice (Nolan & Molla, 2017).

Bernardo, Raul, and Ana participated in the group interviews and shared that their confidence level increased considerably through participating in the summer internship experience. Before the internship, preservice teachers self-reported lower levels of confidence and shared feelings of insecurity toward the teaching profession. The teacher candidates used words such as "so much fear," "so much more stressful," "terrified," or "feeling awkward" when describing their perceptions at the beginning of the internship. After participating in the internship, their confidence level increased considerably. All three students in the focus group used positive language such as "definitely a lot more confident" (Bernardo) or "super confident" (Ana) when describing themselves as teachers.

They all acknowledged that increased confidence adds great value to their teaching practice. All students expressed that they became more excited and committed to teaching as they gained more confidence. Ana reflected as she became more confident, she is more excited to create lesson plans after the internship:

> I'd never seen a lesson plan before the internship, so that [lesson planning] really helped my confidence with creating lesson plans and during step two I was super confident with it and … now I'm really excited (Ana, a female Asian preservice teacher, Interview 2019).

Like Ana, Bernardo described that the internship increased his confidence and that he feels more at ease as a teacher now:

> [I am] definitely a lot more confident because I feel like… I wasn't always confident [before the internship]. But I feel like now I have more of a general confidence and… it's a lot easier being a teacher now (Bernardo, a male Latino preservice teacher, Interview 2019).

The preservice teachers also shared that through the internship, gains in their own teacher confidence improved student engagement and overall student performance. Bernardo shared:

> I feel like if you don't carry yourself in a confident way the kids just won't come to you…I feel like if you are confident in yourself, the efficacy is a lot better, then they will feel comfortable coming to you …and it will in turn help their student achievement. (Bernardo, a male Latino preservice teacher, Interview 2019).

From the preservice teachers' stories, readers can surmise that the ILEs they obtained through the summer internship increased the preservice teachers' confidence, which affected their persistence to further their STEM teaching careers. Kelly (2000) also confirmed that informal learning positively impacts the confidence levels of preservice teachers and their attitudes toward STEM-related subject areas. With respect to being fearful of teaching, the internship program helped shift the attitudes and perceptions of preservice teachers, promoting self-efficacy and increased value in their choice of pursuing the teaching profession. Upon completing the internship, the preservice teachers felt a rise in their level of confidence, which enabled them to better engage with students and enhance student preparation. The confidence in their ability to teach is a strong indicator of teachers' persistence, effort, and work performance (Bandura, 2001).

Strengthened Teacher Motivation and Commitment

As the United States becomes more diverse in racial and ethnic backgrounds, it is imperative that the education sector is strategic in their approach of urban school settings in formal and informal learning environments to cater to the growing diverse populations. Minorities, low Socio-Economic Status (SES), and female students have lower levels of commitment (defined as dedication to STEM in this narrative inquiry study) and motivation (as defined as desire) to persist in STEM (Saw, Chang, & Chan, 2018). Stereotypes unfavorably affect the retention in STEM; however, STEM-related activities (informal and formal experiences) help in the retention of minorities, low SES, and female students (Meador, 2018).

The findings in this study support the notion that the ILEs served as a motivating factor in the preservice teachers' commitment and motivation to persist as STEM majors and in their aspirations to become STEM teachers.

The internship experience positively impacted two of the preservice teachers' desire and dedication to remain and work in STEM. Readers will recall that Bernardo's mom had steered him away from working at a particular neighborhood middle school for his field-based experience. Yet, he persisted with teaching and learned that the was able to motivate students to do their best and that they did, in fact, want to learn. The internship was a pivotal step in solidifying his decision to teach as the experience helped to build his confidence in his teaching abilities.

The perceptions of teaching STEM can affect work–life decisions as seen with Bernardo. Such perceptions may have also served as a factor in Ana's apprehension about teaching STEM. Below Ana recalled how valuable the internship experience was to her in deciding to continue her interests in STEM:

> Yeah, like I first came in and that wasn't the goal. I wanted to see what it was about when I first came to step one. But I think it was mostly the internship. By the end of the internship I was like okay, this is something that I want to do if not forever at least for a little bit (Ana, a female Asian preservice teacher, Interview 2019).

There were direct identifiable experiences gained during the internship that were found in the study. They also are known as the "turning point" experiences that affect preservice teachers' decisions to remain as STEM majors and to pursue teaching as a profession. Ana shared how the internship allowed for building specific skills and practices she had not experienced prior to engaging in ILEs, such as creating lesson plans for the first time: …I'd never seen a lesson plan before the internship, so that [lesson planning] really helped my confidence with creating lessons plans and during step two I was super confident with it… In addition to the effects of ILEs, Raul revealed how the internship experience shared a dynamic feature of consistency with students, which is an experience often not presented in formal learning environments. The interns were able to meet with students daily which helped them in their skills to connect with students in the classroom. Raul shared:

> I mean, I feel like having that experience in general would help just because you're constantly interacting with the kids, kind of like she said, you learn how to talk to them. They [become] comfortable around you… in class versus this [internship experience], you're kind of going back to back…. You really interact with them and get to know them (Raul, a male Latino preservice teacher, Interview 2019).

Like Raul, Bernardo shared an increased commitment to teaching STEM after the internship:

> Yeah, I definitely relate to that because even though I feel like I have learned a lot from this internship, I do want to seek out more like hands-on experiences like this. That way I can get better (Bernardo, a male preservice teacher, Interview 2019).

This particular theme reveals the distinguishable experiences provided by ILEs and the impact of these experiences on scholars' motivation and commitment to STEM teaching as compared to formal learning experiences. During the internship, the motivation and commitment to teach STEM increased for all of the scholars. Nationally, as time progresses, academic initiatives are increasingly allotted less funding or no funding at all in education for student success. As indicated in preservice teacher interviews, explicitly and implicitly, disparities in college access, choice, and readiness impact students' overall scholarly achievement. Raul expressed concern for the education system during the interview, by stating:

> We need to change something because obviously what we're doing right now or what the current state of the education system is right now, something's not working." Ana added to that thought stating "... it makes you like you want to be a part of the change. Like it stops you from being complacent.

While the education system focuses on outputs, the inputs are often not considered especially in the areas of funding and with regard to addressing the needs of underserved and underrepresented students (Roberts et al., 2018). The preservice teachers illuminated evidence that ILEs provide unique learning experiences not often as prevalent in formal education settings.

Improved Classroom Management

Whether novice or seasoned, all teachers have different ways of managing a classroom. However, classroom management techniques should be appropriately used for student engagement and to impact student success. To promote productive learning classroom environments, classroom management helps to convey instructional content focused on the skills of students, as well as their knowledge and abilities (O'Neill & Stephenson, 2011).

For the purpose of this study, classroom management was defined as the inactive and active decisions made by the instructional facilitator for student success. We also found an emerging subtheme in classroom management involving teacher attitudes and behavior, which will be forthcoming in this chapter. The interns in this study were educated in theory and practice approaches for teaching. We examined the participants' ways of managing a classroom by examining the following verbal and nonverbal activities: (1) creating lesson plans, (2) tone of voice, and (3) energy exuded. Overall, the examination of nonverbal and verbal activities, in addition to interviews of study participants, revealed there was a significant increase in the preservice teachers' abilities in classroom management through the informal experiences gained. The student engagement experiences accumulated during the course of the internship helped interns connect and learn how to communicate with students.

Raul learned the importance of communicating with students by relating to them, stating:

> I feel like I've really been able to connect with students and really communicate with them and for me that's really helped out a lot.... How to kind of relate to them. That kind of helps. It helps things move along, especially in the classroom (Raul, a male preservice teacher, Interview 2019).

Like Raul, Bernardo developed a new skill during the internship regarding classroom management by sharing how he engages with students while teaching:

> I feel like I've definitely gotten a lot more effective in [the] classroom because I've learned to read classrooms better (Bernardo, a male preservice teacher, Interview 2019).

As observed in this study, each intern developed their own way to connect with students. In fact, techniques were discovered during the internship that helped students increase their abilities to manage a classroom.

Changed Attitude and Behavior toward Teaching in STEM

The behaviors and attitudes of teachers and students play a dynamic role in student success. For the purpose of this study, behavior is defined as the active and inactive actions expressed. Attitude is defined as the predisposition of the mindset of a person's values, ideas, and thinking patterns. The interns expressed their initial impressions of the students' attitudes toward participation in a summer STEM camp. The interns perceived the students as disinterested in summer learning activities. However, once the interns began interacting with the students during the internship, they experienced a shift in their own attitudes toward the role of teachers in garnering student engagement.

As previously mentioned, the interns shared their perceptions of students' attitudes toward a summer STEM learning experience. Raul mentioned how the students were there involuntarily; however, he took the responsibility of a motivated preservice teacher to inspire students to learn by stating:

> ...their parents made them do, but um, that's not a bad thing. It just depends who, who really is who really up-front teaching them at that point, what are they motivating them to actually be excited about (Raul, a male preservice teacher, Interview 2019).

Ana agreed with Raul about the approach to impact students by sharing:

> ...I do understand that some of the kids didn't necessarily want to be there but I feel like just based on how we did the lessons with them, how we interacted with them... it made it worth it coming in that day. (Anal, a female preservice teacher, Interview 2019).

Bernardo followed Ana's and Raul's interventions to motivate students by echoing:

> ...So I felt like they were just like, oh I'm here because my mom made me come here. And that's when I kind of made it my mission that I need to make a change in someone's life or [have an] impact on them. (Bernardo, a male preservice teacher, Interview 2019).

Because of their shared mindset regarding student demographics, they collectively possessed similar positive attitudes about teaching their students. Raul spoke about the importance of serving as a leader in through modeling behavior to influence students:

> I mean, um, when you're teaching, you kind of have to be the leader of the room. If your leader isn't confident because you know you're cutting the head off the chicken, you know you're going to lose control of the classroom (Raul, a male preservice teacher, Interview 2019).

However, Bernardo shared the energy he exuded was reproduced by students, which sparked student engagement. Here, he described his strategy:

> The students do feed off your energy, so you have to make sure you always kind of seemed confident even though you might be a little bit nervous, you have to make sure you're always upbeat and excited about the lesson because if you walk in you're like, oh, this lesson is so boring. We're just going to get through it. They're going to match the same thing. (Bernardo, a male preservice teacher, Interview 2019)

After their experiences, the intern teachers learned that teaching is more than sharing information and having a positive attitude. There are additional occupational factors that are often not overtly taught to undergraduate students through formal learning experiences. Raul shared how mentoring is a part of the teaching profession by stating:

> I feel like after going through the program…I've learned that teaching is not just about giving information. That's, yeah, that's part of the job of course. But the biggest aspect of it is definitely that mentoring aspect because that's how you really make the difference in a kid's life which is the thing that every teacher should aspire to do. (Raul, a male preservice teacher, Interview 2019).

Bernardo found that there was a counseling aspect to teaching students as well, "So I know I've learned that going along with teaching is that also like a little bit of a counseling aspect (Bernardo, a male preservice teacher, Interview 2019)." Through the ILE, there was a better understanding of the teaching profession. The invaluable opportunities provided by the ILE afforded interns with creative ways for curriculum development. Readers will recall that Ana enthusiastically shared an excitement for creating lesson plans by herself subsequent to creating lesson plans for the internship.

Bernardo, like Ana, was happy to develop a new skill from the internship. However, he incorporated his own personal touch by including culturally relevant pedagogy into his lesson plans. In his response to lesson planning, he stated the following:

> So I felt like I did have a leg up in terms of lesson planning and also learned about culturally relevant pedagogy, how you have to tie your lessons to the culture of the students. That way it feels more personal to them. (Bernardo, a male preservice teacher, Interview 2019).

Raul, the most experienced out of the three interns in the study, explained his level of confidence with lesson planning, "I've always just felt really comfortable making a lesson plan." However, through the internship program he, too, was able to expand not only his abilities to create engaging lessons but also the verbal and nonverbal teaching techniques that encompass classroom management. Ana, for example, shared how the experience shaped her ability to command a classroom, "my teacher voice has definitely gotten louder. It has [gained] a little bit more authority over time." Similarly, Bernardo's also emphasized the need for confidence as they progressed through the program, "I feel like as you go through the internship you do need a little bit of assertiveness." In addition to finding confidence in their classroom presence, the interns also explained the need for adaptability and how the internships helped foster that skill, "I think that's the biggest thing from the internship is learning how to adapt something that doesn't go your way, you know."

IMPLICATIONS AND RECOMMENDATIONS

In the quantitative literature, teacher efficacy has been correlated to student achievement outcomes (Moore & Esselman, 1992; Muijs & Reynolds, 2002), as well as, to student self-efficacy (Anderson et al., 1988; Midgley et al., 1989) (Fig. 1). Through Ana, Raul, and Bernardo's narratives, readers qualitatively learn of the multiple pathways that three STEM undergraduates embarked on to become STEM preservice teachers. Ana and Raul came into college with the initial intention to work in medical fields, while Bernardo had a clear goal of becoming a teacher early on. Their experiences with the *teach*HOUSTON program, especially through the informal internship program, shaped their perceptions of the teaching profession and changed their career trajectories (Ana and Raul) or reinforced their commitment to becoming a STEM teacher (as in the case of Bernardo). Ana, Raul, and Bernardo all wish to make an impact in the lives of their students and challenge the status quo of the American education

Fig. 1. Study Implications.

system through their learning in the informal program. These narrative pieces are sparsely found in the historical and contemporary American and international education literature. Moreover, narrative inquiry is the most appropriate methodology as the stories presented best exemplify the lived experiences of the participants (Clandinin & Rosiek, 2020). The narratives illuminate the critical pathway for Ana, Raul, and Bernardo to become STEM teachers where they will have the chance to impact many students' futures, as well as, contribute to the reform of American public STEM education.

This study was designed to contribute to the literature on teacher preparation programs and help inform university programs regarding the need to include ongoing informal teaching experiences to build not only on pedagogical skills and techniques but also on teacher self-efficacy. The overall increased self-efficacy seen in the study demonstrates the importance of theory and practice in teacher preparation programs. In particular, the Summer STEM Camp and the Noyce Internship Institute increased the self-efficacy of the interns in the following ways: heightened preservice teachers' confidence level, increased motivation and commitment, and improved classroom management. The findings will be of interest to those that would like to enhance their undergraduate teaching preparation programs through offering internships that provide informal learning opportunities through STEM centric outreach efforts such as summer camps. Accordingly, the results of this inquiry help inform our practices as teacher educators to continue providing our preservice teachers with opportunities for professional development through ILEs.

While this study adds to the literature of the effects of ILEs on minority students self-efficacy in teaching secondary education, further research exploration is needed to examine before, concurrent, and after participation in ILEs in relation to the following: content provided in the Noyce Internship Institute, experiences gained through the internship, and outlier factors such as environmental factors (job experience), social networks (parents, peers, instructional supervisors, etc.), or psychological and/or physical conditions of students. The study was fortunate to receive funding through NSF grants, which allowed us to provide access, choice, and readiness in a variety of learning platforms in STEM through formal and informal environments. The needed changes in policies, practices, and theories in education should reflect our findings from this study. Thus, our study provides a compelling example of realizing the dreams, possibilities, and necessity of public education – consistent with other influential preservice teacher programs across America.

AUTHORS' NOTE

This research was funded by the following grants:

- National Science Foundation DUE Award 1240083, *Recruitment, Preparation, and Retention of STEM Students as High School Teachers.*
- National Science Foundation DUE Award 1557309, *University of Houston: Learning through Informal and Formal Experiences.*

- National Science Foundation DUE Award 1557273, *Collaborative Research: Understanding Robert Noyce Teacher Scholarship Outcomes in Texas.*
- National Science Foundation DUE Award 1759454, *Leadership through Equity and Advocacy Development* (LEAD Houston).

REFERENCES

Anderson, R., Green, M., & Loewen, P. (1988). Relationships among teachers' and students' thinking skills, sense of efficacy, and student achievement. *Alberta Journal of Educational Research, 34*(2), 148–165.

Anderson, D., Lawson, B., & Mayer-Smith, J. (2006). Investigating the impact of a practicum experience in an aquarium on pre-service teachers. *Teaching Education, 17*(4), 341–353.

Bandura, A. (1977). Self-efficacy: Toward a unifying theory of behavioral change. *Psychological Review, 84*(2), 191–215.

Bandura, A. (1986). *Social foundations of thought and action: A social cognitive theory*. Englewood Cliffs, NJ: Prentice-Hall.

Bandura, A. (1994). Self-efficacy. In V. S. Ramachaudran (Ed.). *Encyclopedia of human behavior* (Vol. 4, pp. 71–81). New York, NY: Academic Press.

Bandura, A. (Ed.). (1995). *Self-efficacy in changing societies*. New York, NY: Cambridge University Press.

Bandura, A. (1997). *Self-efficacy: The exercise of control.* New York, NY: Freeman.

Bandura, A. (1999). Social cognitive theory of personality. *Handbook of Personality, 2*, 154–196.

Bandura, A. (2001). Social cognitive theory: An agentic perspective. *Annual Review of Psychology, 52*(1), 1–26.

Clandinin, D. J., & Connelly, F. M. (2000). *Narrative inquiry: Experience and story in qualitative research* San Francisco, CA: Jossey-Bass.

Clandinin, D. J., & Rosiek, J. (2020). *Journeys in narrative inquiry: The selected works of D. Jean Clandinin*. London: Routledge.

Connelly, F. M., & Clandinin, D. J. (1990). Stories of experience and narrative inquiry. *Educational Researcher, 19*(5), 2–14.

Craig, C. (2011). Narrative inquiry in teaching and teacher education. In J. Kitchen, D. Parker, & D. Pushor (Eds.), *Narrative inquiries into curriculum making in teacher education* (pp. 19–43). Bingley: Emerald Publishing Limited.

Craig, C., Evans, P., Verma, R., Stokes, D., & Li, J. (2019). A tribute to "unsung teachers": Teachers' influences on students enrolling in STEM programs with the intent of entering STEM careers. *European Journal of Teacher Education, 42*(3), 335–358.

Craig, C., You, J., Zou, Y., Verma, R., Stokes, D., Evans, P., & Curtis, G. (2018). The embodied nature of narrative knowledge: A cross-study analysis of embodied knowledge in teaching, learning, and life. *Teaching and Teacher Education, 71*, 329–340.

Darling-Hammond, L. (2012). *Powerful teacher education: Lessons from exemplary programs*. New York, NY: John Wiley & Sons.

Denson, C., Austin, C., Hailey, C., & Householder, D. (2015). Benefits of informal learning environments: A focused examination of STEM-based program environments. *Journal of STEM Education, 16*(1), 11–15.

Dewey, J. (1938). The pattern of inquiry. *The Essential Dewey, 2*, 169–179.

DeWitt, J., & Storksdieck, M. (2008). A short review of school field trips: Key findings from the past and implications for the future. *Visitor studies, 11*(2), 181–197.

Dierking, L. D., Falk, J. H., Rennie, L., Anderson, D., & Ellenbogen, K. (2003). Policy statement of the "informal science education" ad hoc committee. *Journal of Research in Science Teaching, 40*(2), 108–111.

Gauthier, P. (2010). Confidence [Editorial]. *Gastroenterology Nursing, 33*(3), 186–187. doi:10.1097/SGA.0b013e3181e82c4b

Gerber, B. L., Cavallo, A. M., & Marek, E. A. (2001). Relationships among informal learning environments, teaching procedures and scientific reasoning ability. *International Journal of Science Education*, *23*(5), 535–549.

Gibson, S., & Dembo, M. H. (1984). Teacher efficacy: A construct validation. *Journal of Educational Psychology*, *76*(4), 569–582.

Guskey, T. R., & Passaro, P. D. (1994). Teacher efficacy: A study of construct dimensions. *American Educational Research Journal*, *31*(3), 627–643.

Harlow, D. B. (2012). The excitement and wonder of teaching science: What pre-service teachers learn from facilitating family science night centers. *Journal of Science Teacher Education*, *23*(2), 199–220.

Jackson, C., Mohr-Schroeder, M., & Little, D. L., II. (2014). Using informal learning environments to prepare preservice teachers. *Teacher Education & Practice*, *27*(2–3), 445–463.

Johnson, D., & Chandler, F. (2009). Pre-service teachers' fieldtrip to the battleship: Teaching and learning mathematics through an informal learning experience. *Issues in the Undergraduate Mathematics Preparation of School Teachers: The Journal*, *2*, 1–9.

Kelly, J. (2000). Rethinking the elementary science methods course: A case for content, pedagogy, and informal science education. *International Journal of Science Education*, *22*(7), 755–777.

Kotys-Schwartz, D., Besterfield-Sacre, M., & Shuman, L. (2011, October). Informal learning in engineering education: Where we are—Where we need to go. 2011 Frontiers in Education Conference (FIE), pp. T4J-1. IEEE.

Martin, L. M. (2004). An emerging research framework for studying informal learning and schools. *Science Education*, *88*(S1), S71–S82.

Meador, A. (2018). Examining recruitment and retention factors for minority STEM majors through a stereotype threat lens. *School Science & Mathematics*, *118*(1–2), 61–69.

Midgley, C., Feldlaufer, H., & Eccles, J. S. (1989). Student/teacher relations and attitudes toward mathematics before and after the transition to junior high school. *Child development*, 981–992.

Mohr-Schroeder, M. J., Jackson, C., Miller, M., Walcott, B., Little, D. L., Speler, L., … Schroeder, D. C. (2014). Developing middle school students' interests in STEM via summer learning experiences: See Blue STEM Camp. *School Science & Mathematics*, *114*(6), 291–301.

Moore, W. P., & Esselman, M. E. (1992). Teacher efficacy, empowerment, and a focused instructional climate: Does student achievement benefit? Paper presented at the annual conference of American Educational Research Association. San Francisco, CA.

Muijs, D., & Reynolds, D. (2002). Teachers' beliefs and behaviors: What really matters?. *Journal of Classroom Interaction*, 3–15.

National Academy of Sciences. (2007). *Rising above the gathering storm: Energizing and employing America for a brighter economic future* Washington, DC: National Academy Press.

National Research Council. (2005). America's lab report: Investigations in high school science.

Newell, A. D., Zientek, L. R., Tharp, B. Z., Vogt, G. L., & Moreno, N. P. (2015). Students' attitudes toward science as predictors of gains on student content knowledge: Benefits of an after-school program. *School Science & Mathematics*, *115*(5), 216–225. doi:10.1111/ssm.12125

Nolan, A., & Molla, T. (2017). Teacher confidence and professional capital. *Teaching and Teacher Education*, *62*, 10–18.

O'Neill, S. C., & Stephenson, J. (2011). Classroom behaviour management preparation in undergraduate primary teacher education in Australia: A web-based investigation. *Australian Journal of Teacher Education*, *36*(10), 35–52.

Olson, J. K., Cox-Petersen, A. M., & McComas, W. F. (2001). The inclusion of informal environments in science teacher preparation. *Journal of Science Teacher Education*, *12*(3), 155–173.

Pajares, F. (1996). Self-efficacy beliefs in academic settings. *Review of Educational Research*, *66*(4), 543–578.

Phillips, M., Finkelstein, D., & Wever-Frerichs, S. (2007). School site to museum floor: How informal science institutions work with schools. *International Journal of Science Education*, *29*(12), 1489–1507.

Polkinghorne, D. (1995). Narrative configuration in qualitative analysis. In J. A. Hatch & R. Winiewski (Eds.), *Life history and narrative*. Washington, DC: The Falmer Press.

Popovic, G., & Lederman, J. S. (2015). Implications of informal education experiences for mathematics teachers' ability to make connections beyond formal classroom. *School Science & Mathematics, 115*(3), 129–140.

Putnam, R. T., & Borko, H. (2000). What do new views of knowledge and thinking have to say about research on teacher learning? *Educational Researcher, 29*(1), 4–15.

Rew, W. J. (2013). Instructional leadership practices and teacher efficacy beliefs: Cross-national evidence from Talis. Retrieved from http://purl.flvc.org/fsu/fd/FSU_migr_etd-7573

Roberts, T., Jackson, C., Mohr-Schroeder, M. J., Bush, S. B., Maiorca, C., Cavalcanti, M., … Cremeans, C. (2018). Students' perceptions of STEM learning after participating in a summer informal learning experience. *International journal of STEM education, 5*(1), 35.

Saw, G., Chang, C. N., & Chan, H. Y. (2018). Cross-sectional and longitudinal disparities in STEM career aspirations at the intersection of gender, race/ethnicity, and socioeconomic status. *Educational Researcher, 47*(8), 525–531.

Siwatu, K. O. (2007). Preservice teachers' culturally responsive teaching self-efficacy and outcome expectancy beliefs. *Teaching and Teacher Education, 23*(7), 1086–1101.

Skaalvik, E. M., & Skaalvik, S. (2007). Dimensions of teacher self-efficacy and relations with strain factors, perceived collective teacher efficacy, and teacher burnout. *Journal of Educational Psychology, 99*(3), 611–625.

Stokes, D., Evans, P., & Craig, C. (2017). Developing STEM teachers through both informal and formal learning experiences. Ediciones Universidad de Salamanca.

Sullenger, K. (2006). Beyond school walls: Informal education and the culture of science. *Education Canada, 46*(3), 15–18.

Tschannen-Moran, M., & Hoy, A. W. (2001). Teacher efficacy: Capturing an elusive construct. *Teaching and Teacher Education, 17*, 783–805.

Tuchman, E., & Isaacs, J. (2011). The Influence of formal and informal formative preservice experiences on teacher self-efficacy. *Educational Psychology, 31*(4), 413–433.

Wallace, C. S., & Brooks, L. (2014). Learning to teach elementary science in an experiential, informal context: Culture, learning, and identity. *Science Education, 99*(1), 174–198.

Windschitl, M. (2009). *Cultivating 21st century skills in science learners: How systems of teacher preparation and professional development will have to evolve*, Presentation given at the national academies of science workshop on 21st century skills. Washington, DC, Vol. 15.

Wolters, C. A., & Daugherty, S. G. (2007). Goal structures and teachers' sense of efficacy: Their relation and association to teaching experience and academic level. *Journal of Educational Psychology, 99*(1), 181.

Yilmaz, M., Ren, J., Custer, S., & Coleman, J. (2010). Hands-on summer camp to attract K–12 students to engineering fields. *IEEE Transactions on Education, 53*(1), 144–151. doi:10.1109/te.2009.2026366

Zientek, L. R., & Thompson, B. (2008). Preparing high quality mathematics and science teachers: Are we meeting the challenge? *Research in the Schools, 15*, 1–19.

PARENTS' INFLUENCE ON UNDERGRADUATE AND GRADUATE STUDENTS' ENTERING THE STEM DISCIPLINES AND STEM CAREERS

Cheryl J. Craig, Rakesh Verma, Donna W. Stokes, Paige K. Evans and Bobby Abrol

ABSTRACT

This research examines the influence of parents on students studying the science, technology, engineering, and mathematics (STEM) disciplines and entering STEM careers. Participating youths were awarded scholarships from large funded US grant programmes. Cases of two graduate students (one female, one male) and one undergraduate student (male) are featured. The first two students in the convenience sample are biology and physics majors in a STEM teacher education program; the third is enrolled in computer science. National reports emphasizing the importance of parents on their children's education are presented, along with diverse international literature. The use of narrative in STEM curriculum and narrative inquiry in STEM research are also documented. Experience, story, and identity form the study's conceptual frame. The narrative inquiry research method employs broadening, burrowing, and storying and restorying to elucidate the students' academic trajectories. Incidents of circumstantial and planned parent curriculum making surfaced when the data were serially interpreted. Other noteworthy themes included: (1) relationships between (student) learners and (teacher) parents, (2) invitations to inquiry, (3) modes of inquiry, (4) the improbability of certainty, and (5) changed narratives = changed lives. While policy briefs provide

Preparing Teachers to Teach the STEM Disciplines in America's Urban Schools
Advances in Research on Teaching, Volume 35, 109–133

ISSN: 1479-3687/doi:10.1108/S1479-368720210000035007

sweeping statements about parents' positive effects on their children, narrative inquiries such as this one illuminate parents' inquiry moves within home environments. These actions became retrospectively revealed in their adult children's lived narratives. These small stories, while not generalizable, map how students, shaped by their parents' nurturing, enter the STEM disciplines and STEM-related careers through multiple pathways in addition to the identified pipeline.

Keywords: Parent involvement; STEM education; STEM pipeline; STEM pathway; science as inquiry; physics as inquiry; computer science education; narrative inquiry; curriculum making; *teach*HOUSTON

While the public generally agrees that science, technology, engineering, and mathematics (STEM) are critically important to economic prosperity (e.g., President's Council of Advisors on Science and Technology, 2010), not enough students are entering the STEM disciplines to satisfy present and future workforce needs. Furthermore, the percentages of females and those representing minority populations are proportionately low. This serious situation has led the National Science Foundation (NSF) in the United States and other agencies internationally to award scholarships to encourage undergraduate and graduate students to study STEM-related subject matter. These scholarships are meant to address "leakage" (Cannady, Greenwald, & Harris, 2014, p. 444) in the STEM pipeline (e.g., National Academy of Science, 2010) that occurs between and among high school, undergraduate education, graduate studies, and STEM-related careers.

For Cannady et al. (2014), the pipeline metaphor inadequately problematizes why students are not entering STEM fields and assuming STEM positions as expected. First, the image of a pipeline implies "a lock-step trajectory that fails to account for alternate pathways to STEM careers" (p. 448). Second, contextual factors that shape the "long and winding path" toward STEM study and employment become lost in the metaphor (p. 445). Third, "…middling results call into question the usefulness of the pipeline metaphor as a framework for policy decisions and inquiry into the issue" (p. 444). Consequently, the authors prefer the multiple pathways metaphor to the narrowing pipeline image as "it invites inquiry into the journey rather than simply headcounts at the mile markers" (p. 455).

One recognized, but underexplored, phenomenon relating to the STEM cradle-to-career trajectory is the impact of family. In the *Age of STEM*, Freeman, Marginson, and Tytler (2014) state that most, if not all, countries report that parents are important because they affect "whether positive attitudes to STEM lead to STEM study and work" (p. 4). This suggests that STEM identity imprinting may happen in families (especially Asian ones) before youths begin formal schooling (Nash, 2014a, p. 40).

This chapter focuses on the influence of parents on undergraduate and graduate STEM students who received NSF scholarships to become biology/

chemistry/physics teachers or to enter into the computer science field at an American research-intensive university. In our exploratory research study, a senior computer science professor was the first to alert us to the key roles that parents play. That Asian American male stressed "the need to reach parents" as a pivotal factor in addressing the shortage of STEM graduates entering the workforce. What he said confirmed what many undergraduate and graduate students had already revealed in interview and focus group settings. This chapter captures what we learned about the influence of parents through observing STEM students and faculty members, engaging them in interview and focus group conversations, and videotaping them in action in classrooms and laboratories.

LITERATURE REVIEW

Three strands of literature situate this research: (1) STEM in school curriculum, (2) the influence of parents on student success, and (3) the use of narrative in STEM curriculum and in STEM research.

STEM in School Curriculum

How STEM is formally taught along the educational continuum is disputed. Existing issues pertain to inquiry (Crawford, 2000), integration (National Research Council, 2014), constructivism (Driver, Asoko, Leach, Scott, & Mortimer, 1994), conceptual change (Duit & Treagust, 2003), the politics of knowledge (Brown, 2017), relevance (Stuckey, Hofstein, Mamlok-Naaman, & Eilks, 2013), standards (Wallace, 2012), assessment (Bybee, 2013), accountability (National Research Council, 2013), and pedagogical content knowledge (Van Driel & Berry, 2012), among a bevy of other concerns, including the nature of science.

Amid this contestation, the teaching of STEM unfurls through a universally accepted process (Goodson, 1990): curriculum making (Schwab, 1973). Schwab determined that every curriculum making situation requires a teacher, a learner, subject matter, and milieu, with all being of equal value (Fig. 1). Regardless of age, youths are individual learners essential to the curriculum making act.

Recently, Schwab's curriculum making framework has been used to address the influence of teachers on high school students' intentions to declare STEM majors in college (Bottia, Stearns, Mickelson, Moller, & Parker, 2015). Also, Huber, Murphy, and Clandinin (2011) reconceptualized those represented in Schwab's commonplaces of curriculum (Fig. 2). These authors wondered: "Could familial curriculum making be understood as equally important to the negotiation of [meaning] as in-classroom, in-school curriculum making?" In other words, "could a parent be the teacher?" Could the concept of education "growing human beings" (Bobbitt, 1924/1975, p. 17) happen both inside and outside of schools and universities?

In their research study, Huber et al. (2011) confirmed that familial curriculum making occurs. Moreover, this new notion of a cocomposed curriculum verified

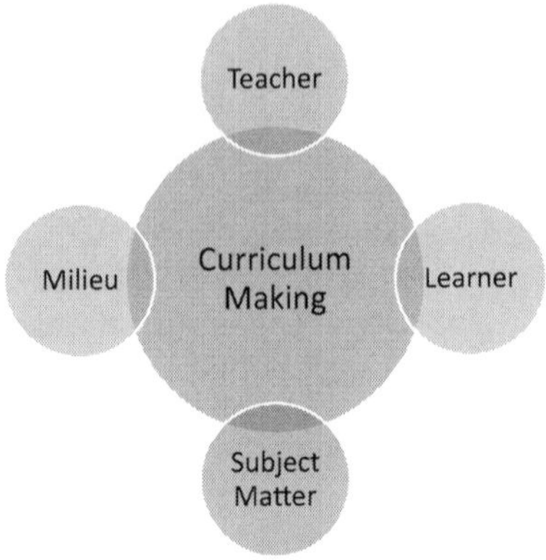

Fig. 1. Schwab's Commonplaces of Curriculum.

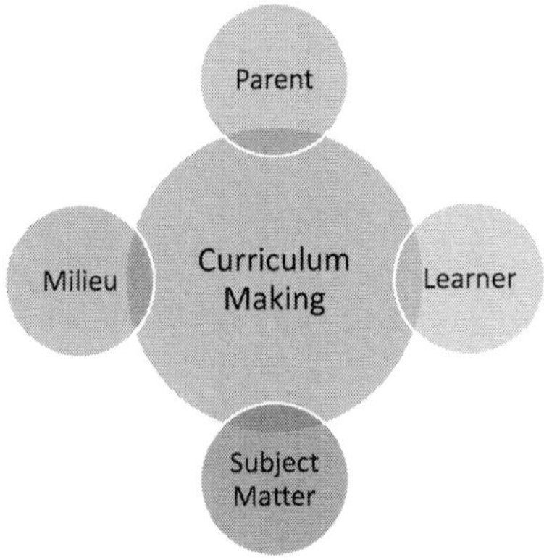

Fig. 2. Familial Commonplaces of Curriculum.

that youths can concurrently live in more than one curriculum making place (home *and* school) and that curriculum is indeed a "multistoried process" (Olson, 2000a). For us, this reinforced that the focus of our research inquiry was on target.

Influence of Parents on Student Success

We now present our second strand of the literature: how parents contribute to student success. Both American policy documents and national and international literature bases will be presented.

US Research

Two prominent American documents, the Coleman Report (Coleman et al., 1966) and the Jencks Report (Jencks et al., 1972), along with contemporary literature (e.g., Ratelle, Larose, Guay, & Senecal, 2005), have upheld that family impacts student achievement.

The Coleman Report

The Coleman Report, the most important twentieth-century American policy document, laid the foundation for the educational course of action in all subject areas during the civil rights era. While some elements of Coleman et al.'s report were controversial, one finding has managed to survive the test of time: the positive influence of family on student success. This finding, unfortunately, was descriptive, not evidential. More than a half-decade later, Hanushek (2016) deemed the family effect "the bumper-sticker claim of those [citing] the Coleman Report." He furthermore confirmed that little was known in Coleman's era or is known presently about the ways that family effects play out. In Hanushek's words, "the exact channels through which family…impacts…educational and lifetime successes remain uncertain."

The Jencks Report

The Jencks Report (Jencks et al., 1972) closely followed the Coleman Report. Using Coleman's datasets supplemented with new survey results, Jencks et al. confirmed that family has a major impact on students' cognitive skills. While Jencks et al. are recognized for determining why some students succeed and others do not, his research team once again did little to unpack how family members cognitively affect offspring and sibling achievement. The Jencks Report also generalized the influence of family with limited data to support its finding.

Contemporary American Research

Since the two influential reports, STEM research has repeatedly confirmed the positive impact of family and parents on student success in the United States. This is evident in work disseminated by the National Center for Family and Community Connections with Schools, Southwest Educational Development Laboratory, the National Academy, and the National Middle School Association. The most promising, contemporary research is a longitudinal (35-year) research report recently released by the National Academy. Conducted by Rozek, Svoboda, Harackiewicz, Hulleman, and Hyde (2017), the study builds on an animal behavior investigation in psychology rather than on historical policy documents. The research team found that parents make a difference because they cultivate their children's STEM interests and communicate the importance of STEM to society. Rozek et al. termed their finding an "indirect pathway," which refutes the pipeline imagery they originally used to position their study in the literature.

International Research

The European Union–funded project, *Interest and Recruitment in Science (IRIS)*, is the most significant international study about the in-school and out-of-school influences that attract students to STEM (Henricksen, Dillon, & Ryder,

2015). One major outcome is the proven effects of parents on youths entering the STEM disciplines and embarking on STEM careers. Also, International Journal of Science Education (*IJSE*) carried an article by Cleaves (2005) that addressed students' formations of secondary science choices. Among its findings was the impact of "significant others" on students' choices. The term, "significant others," additionally suggests a multiple pathway approach to STEM rather than a narrow pipeline.

Joining our survey of the literature's first two strands is our third strand: the use of narrative in STEM education and narrative research in STEM education.

Use of Narrative in STEM Curriculum and Narrative Research in STEM Education

Narrative and narrative inquiry have been used in nearly every field from philosophy and psychoanalysis to organizational theory and computational mathematics. To Schwartz (2015), story is the "secret sauce" on which the STEM disciplines have not fully capitalized. Two principle ways that narrative is used in STEM will now be discussed: narrative in STEM curriculum and narrative research in STEM education.

Narrative in STEM Curriculum

Because STEM has "no fixed catalog of disciplines" from which structure is derived (Schwab, 1961/1978), scientific change happens when humanly created paradigms of knowledge shift (Kuhn, 1970/1996). Reduced to their essence, accepted bodies of knowledge change when humanly constructed narratives change, as Nash (2014b) explains:

> Without...stories ...lives are without form or content. So too with subject matter. When all the technical trappings are removed, an academic discipline is really a particular group's storied view of the world, and the story changes often. Find the story of the subject matter...and you have located the core of the academic discipline. (p. 30)

Thus, teaching science as if it were a story and/or through stories is what Conle (2003) proposed in her "anatomy of narrative curricula" and what Egan (1985) meant when he described "teaching as storytelling." It is also what Olson (2000a) referred to in her conceptualization of curriculum as a "multistoried process." Blackley and Howell's (2015), *A STEM narrative: 15 years in the making*, forms a prime international example of this approach.

Narrative Research in STEM Education

A second way that narrative is used in STEM is to study STEM subjects using narrative research methods. A narrative inquiry in teaching and teacher education literature review has been conducted that captured the state of the field until the early 2000s (Craig, 2011). Since then, narrative inquiry research has been undertaken by a mathematics and mathematics education faculty at a minority serving university (Sack, Quander, Redl, & Leveille, 2016), in a secondary STEM teacher education program (Craig et al., 2017a, 2017b), in cybertechnology (Craig et al. final decision), in a mathematics teacher education program

(Kaasila, 2007), and in an experimental environmental education project (Hwang, 2011). Additional narrative research has been conducted in science teacher professional development (Van Driel, Beijaard, & Verloop, 2001), complexity science (Luhman & Boje, 2001), video-case research (Derry et al., 2010), one-to-one computing (Fleischer, 2012), mathematics education (Lerman, 2000), mathematics reform (Drake & Sherin, 2006), engineering education (Case & Light, 2011), and feminist engineering (Beddoes, 2013). As can be seen, the use of narrative research methods is gaining momentum in all disciplines in the STEM field. However, to this point in time, it has not been employed to research the influence of parents on undergraduate and graduate students' entering the STEM disciplines and STEM careers, the focus of this chapter.

CONCEPTUAL FRAMEWORK

Three concepts underpin this narrative inquiry into parents' shaping effects on their children's STEM choices and career paths: experience, story, and identity.

Experience

Life is experience, and education likewise is experience (Dewey, 1938). When life commences, experience and education begin, too. Students' experiences form an integral part of their lives from birth to high school to university and beyond. To Dewey, each new experience takes up something new from the previous experience, informing it in instructive or noninstructive ways. Thus, experience has both active and passive elements. When students engage in a science experiment, for example, they act it out while simultaneously living with its consequences (Dewey, 1916, p. 163). Furthermore, their current science experiences build on past science experiences, which inform their future science experiences (Dewey, 1938). On this continuum, experience cumulatively shapes how students live science, how they learn science, and ultimately, whether science becomes a part of who they are. This point is important because we center on students' retrospective experiences of their parents introducing them to STEM as youngsters and follow the expansion of students' realms of experience over two-to-four-year periods.

Story

Experience is intimately connected to narrative. Experience can only be captured through storytelling and represented narratively. Reflection is similarly implicated because storying and restorying involves probing changes and continuities through looking backward and forward, inside and out (Clandinin & Connelly, 2000). "Following where [stories] lead" (Olson, 2000b) is how human lives take shape, particularly where STEM experiences and experiments are concerned.

To Bruner (2002), "lives with stories start early and continue ceaselessly" (p. 1). Children enter the world narratively through responding to their parents' cooing. Families begin to educate youngsters by sharing stories (possibly about science?) when children are "as blank and unresisting as [they] are ever going to be" (Stone, 1988, p. 10). Through story exchanges, "a child's education is instantly, naturally, and authentically underway" (Pushor, 2013, p. 7). It follows then that youth grow with certain narratives "under their skin." Some tales are "weightless pleasures"; others are negatively imprinted on their psyches (Stone, 1988, p. 10). Family stories mostly communicate the ways of the world. Family members are so influential that the family unit has been described as a child's "first culture" (Stone, 1988, pp. 6–7). Families make sense of the world in the same way a second language is learned (Bellah, Madsen, Sullivan, Swidler, & Tipton, 2007). Family narratives "model not only a world, but minds seeking to give [the world] meaning" (Bruner, 2010, p. 27). Through sharing personal narratives, the world embodied in the stories people live and tell becomes known.

Until memory develops, parents control the plotlines of their children's narratives (Bauer & Fivush, 1992; Fivush & Buckner, 2003; Fivush & Nelson, 2004). However, as youths mature, these storylines decrease in importance. Youths begin to compose their own "stories to live by" (Connelly & Clandinin, 1999). In Stone's words, "[children] come of age and tell the stories of [their] lives in which their past[s] become the prologue to their futures." For Stone, young adults achieve balance through mediating this myriad of past-present-future stories. They develop "a way to be part of and apart from [their] families"; they discover "way[s] of holding on and [ways of] letting go" (Stone, 1988, p. 244).

Identity

Because family is integral to culture, it is the "staging ground" (Maxwell, 2001, p. 1) for students' "stories to live by." "Stories to live by" refer to "our identities [being] composed and improvised as we go about living our lives embodying knowledge and engaging our contexts" (Connelly & Clandinin, 1999, p. 4). "Stories to live by" are both personal and social. Students' "stories to live by" are shaped by their knowledge forged in multiple contexts. Concurrently, students' "stories to live by" are affected by the landscapes in which they have lived and the feedback they have received (Clandinin & Connelly, 1995). What they hear from others (about science?) "give[s] messages and instructions; …offer[s] blueprints and ideals; [and]… issue[s] warnings and prohibitions" (Stone, 1988, p. 5), which, in turn, influences their identities. In this way, identities are repeatedly "formed and reformed by the [multitude of] stories [they are told] and which [they] draw upon in [their] communications with others" (Beijaard, Meijer, & Verloop, 2004, p. 123). Included in this swirl of experiences are students' developing STEM identities (Carlone & Johnson, 2007; Chang, Eagan, Lin, & Hurtado, 2011; Hazari, Sadler, & Sonnert, 2013) and the different patterns of parent support they have experienced (Aschbacher, Li, & Roth, 2010).

Within this ever-expanding constellation of stories (Craig, 2007), family members continue to be important shaping forces on students' identities (Jackson & Seiler, 2013). Family "fasten[s] identity in place and keep[s] it from floating off… or losing its shape…" (Stone, 1988, p. 11). Concomitantly, if youths wish to change their identities, they can alter the plotlines of the stories they live and tell, and transform the courses of their lives (King, 2003).

We now launch into a description of our research methodology, which is followed by the narratives of experiences of the three STEM university students we have chosen to feature: Katrina who took the traditional path – the pipeline route – to STEM education and employment, and Ryan and Sam who traveled long and winding roads and found alternate pathways to become STEM professionals.

RESEARCH METHOD

Qualitative research methods in STEM education capture how events unfurl, identify subjective sense-making spurring action, and assist in the understanding of how change happens (Erickson, 2012, p. 1451). Narrative inquiry is a qualitative method that allows for multiperspectival views of science, one that speaks less authoritatively than ethnography (Erickson, 2012, p. 1467). Like scientific inquiry, "[most qualitative methods] require…the…mind to be open to the unknown, and as empty, as much as possible, from preconceived ideas" (Viray, 2014, p. 190). It is critically important that the knowledge paradigm underlying one's qualitative research method fits the STEM phenomenon under study (Taylor & Wallace, 2007).

Introduction to Narrative Inquiry

Reduced to its essence, narrative inquiry is the "experiential study of experience" (Xu & Connelly, 2009, p. 354), with narrative being the phenomenon of inquiry *and* the method of investigation (Connelly & Clandinin, 1990). Story is "the portal through which a person enters the world" (Connelly & Clandinin, 2005, p. 477); it is the "portal to experience" (Xu & Connelly, 2009, p. 352). Narrative inquiry avoids strategies, rules, and techniques (Clandinin & Connelly, 2000) because "life is not made up of separate pieces" (Bateson, 1994, p. 108). The experience of teaching and learning science are personal and emotional as well as cognitive and rational (Hollingsworth, Dybdahl, & Minarik, 1993). Hence, narrative inquiry is

> …not so much [about] a structured answer to a [STEM-related] question, or a way for accounting for actions and events, as it is a gateway (a pathway?)… into meaning and significance. (Xu & Connelly, 2009, p. 356).

Narrative inquiry has three commonplaces. The term, commonplace, connects the method's science as inquiry roots to Schwab (1969), whereas temporality, sociality, and place/situation locate narrative inquiry's origins in Dewey's (1938) experiential philosophy. Thus, to reflect narrative inquiry's field-based lineage, all three commonplaces must be present and simultaneously interacting in studies embracing this research genre.

Sources of Evidence

Four forms of evidence provide this chapter's narrative grist. Three are interviews, focus groups, and videotaped classwork and interactions; the fourth is reflective journal writing or email communications. This labeled and dated evidence formed our research record.

Tools of Analysis

Three analytical tools (Connelly & Clandinin, 1990) are used to report this study's storied evidence. Through *broadening*, our work traces to the Cold War, the United States's national security need to be on the forefront of scientific discovery and the founding of the NSF. The Soviet moon landing of Sputnik in 1957 and *A Nation at Risk* (Gardner, 1983) further increased the importance of NSF. Funded projects like our own are products of NSF's history. Through *burrowing*, our second tool, we focus on the stories of three undergraduate/graduate students. The third analytical tool, *storying and restorying*, helps us to animate familial curriculum making and to elucidate pipeline and multiple pathway entries into STEM study and STEM employment.

To these devices, *serial interpretation* (Schwab, 1954/1978) has been added to spark discussions about "encompassing idea[s]" that transcend multiple studies. Serial interpretation involves "talking across" (Stone, 1988, p. 2) research projects, rather than reporting a single case. To Schwab, serial interpretation adds heft and power to small studies. Serial interpretation enabled us to produce both pipeline *and* pathway exemplars, as readers will soon see.

Trustworthiness of Findings

Qualitative research does not seek a single answer or a single truth to a query. Instead, narrative inquiry's texts convey truthlikeness (Bruner, 1987), which is what comprises the trustworthiness (Mishler, 1990) of our findings. To strengthen our study's trustworthiness, we conducted member checks with our participants and traced truth claims to a minimum of three sources of evidence. For example, when we included that a senior computer science professor suggested that we should pay more attention to parents, we had five sources of evidence entered into the research record supporting his claim: (1) a videotape of his interview, (2) a transcript of his interview, (3) the first author's research journal entry, (4) participant observation notes written by one research assistant, and (5) participant observation notes written by a second research assistant. Data were collected for our research participants over a four-year period (STEM teacher education program) and a two-year period (computer science program). In all cases, "narrative resonances" (Conle, 1996) were sought between what was voiced by the participating students and professors and what we as research team members observed, read, videotaped, and chronicled in our research record.

Overall, narrative inquiry research revolves around narrative truth (Spence, 1984). It follows where stories lead rather than being driven by time-ordered historical facts. We do not wish to suggest that narrative inquirers are oblivious to historical truth. The bottom line is that we entered this STEM study narratively rather than historically. Hence, we present our research texts narratively rather than historically. In the end, though, readers decide whether this research study concerning the influence of parents informs their knowing and contains semblances of truth that are actionable (Lyons & LaBoskey, 2002).

Having explained our research methodology, we now introduce the undergraduate and graduate STEM students whose narratives of experience we spotlight.

Description of Undergraduate and Graduate Students

Although most students in the grant programmes (N = 35 to date) recalled the influence of their parents and families on their early learning, we decided to focus on a convenience sample of three.[1] In accordance with our human subjects' agreements, pseudonyms have been used instead of their real names. "Katrina Roderick" and "Ryan Archer" were part of the NSF Noyce grant program; "Sam Bernard" was in the NSF Department of Undergraduate Education grant program.

We chose Katrina, Ryan, and Sam from our data pool because they presented contrasting academic trajectories. Katrina seamlessly finished high school, entered university, completed her program of studies, and became a teacher. Ryan and Sam embarked on different journeys. Ryan completed high school, entered and left university, and returned two decades later. As for Sam, he obtained his general educational development (GED) certificate (an alternate way to complete high school), completed two years at a college, and then transferred his coursework to the research-intensive university.

Katrina Roderick

Katrina Roderick came from a family of five children, all with university degrees. In high school, she enrolled in advanced placement biology. At university, she majored in biology and minored in teaching. Katrina Roderick teaches high school biology and advanced placement science classes in an adjoining state. To Katrina, her dream job would have two foci: (1) field work in biology; and (2) master teacher in a teacher development program like physics educator, Paige Evans (fourth author).

Ryan Archer

As a youngster, Ryan Archer had learning needs rarely met through formal schooling, despite him being gifted-and-talented and being enrolled in advanced placement classes. Ryan began university at 16, but dropped out to work as a regional director of employee development at a technology store. That experience financially leveraged his return to university in his 30s. Ryan teaches high school physics in an underserved high school. His ideal position would be to become a physics teacher educator and prepare teachers also like physics educator, Paige Evans (fourth author).

Sam Bernard

Sam Bernard came from two blended families of seven children. Sam's father owns and operates an air-conditioning company. His mother, who is now deceased, worked as a bookkeeper. Sam has one older sister, four younger brothers, and one younger sister. His older stepsister has a social work degree. His full-blooded brother dropped out of community college to become an instructional technology consultant. Sam's remaining stepsiblings are still in K–12 schools. Although his parents divorced when he was three and he lived with his mother, he maintained contact with his father. Sam did not attend high school due to "social anxiety." Instead, he enrolled in a preparatory GED program and then dropped out when his mother, his brother, and he were evicted from their apartment. Shortly thereafter, he successfully received his GED document. Sam then entered community college where he enrolled in remedial courses that allowed him to transfer credits to a university. Sam, who is presently a computer programmer in a midsized business, would ideally found a start-up company that would address the demand for innovation in the computer science industry.

NARRATIVES OF EXPERIENCE

Katrina Roderick's Account

Katrina Roderick connected her science roots to being born the daughter of "a dad who is a scientist" – "a person who is naturally into science." Katrina explained that her father:

> …always wanted to do greater things [with his life], but he had children young…And so he ended up being a registered nurse…But he does a lot of [science] on his own… He loves learning…He's a genius…He's crazy…

According to Katrina, her father set up chemistry and physics experiments in the family's kitchen almost every weekend for her siblings and her. These experiments included inflating balloons with vinegar and baking soda, balancing the broom acts, bending water without touching it, and learning about thermal energy transfer through baking cakes and cupcakes. One experiment involved taste-testing, which was accompanied by data collection and analysis. Katrina said that her dad "came home with three or four different types of hot dog relish. We sat and shared our data after trying each brand and comparing our notes." She added: "…. I can't even begin to count the times that I awoke and did… experiments with him…"

Kitchen science was not the only activity in which Katrina's family participated. Her father also engaged his children in night-time inquiries. One activity involved observing and naming the different constellations of stars. Katrina shared what happened:

> We would go into the backyard and lay on…blankets… He would point out the different clusters of stars and give them names. He would ask which ones we remembered (usually not be too many!!!) and then he would talk about how stars change, what they are, why some are brighter and so on.

Overall, Katrina said her father taught us that we "need to accept that we will never know everything and that it is okay not to know…" Concomitantly, her father would require everyone to outline how they would resolve their unanswered queries. He would also "ask them to EXPLAIN (capitals in original) how we knew our answers were correct." Katrina reflected: "Honestly, this was his most important lesson – teaching us how to learn." She continued:

> He helped build a mindset of curiosity through questioning assumptions and providing information to help me answer my questions. [Concurrently] it was never okay [with him] to remain ignorant. I use this [family learning] every day… This idea of being able to accept what I know and explore what I don't know… It is okay to be wrong [if] I am in pursuit of an answer.

It is little wonder, then, that Katrina is explicit about her father being her "inspiration for anything I do that is science because he gets so excited about it. Science is his muse."

The lifelong paternal support Katrina received channeled her into the science education field. Her father spoke often about her becoming a teacher. To her, his molding of her naturally "mov[ed] her in the direction of science."

Because Katrina wanted to obtain a composite teaching credential, she was one of the first students to enroll in a physics as inquiry course.[2] According to Katrina, the process of inquiry captured her interests because it reminded her of childhood experiments with her father:

> It was not following a workbook. It was not "do this," "do that." It was "let's try to figure out how to do this…It was exciting. I became pumped by science taught as inquiry…I had so many ah-hah moments… I would get so excited…."

Katrina's excitement in her physics teacher education classroom mirrored her father's enthusiasm for scientific inquiry. Katrina similarly became just as "enthralled" by science as inquiry. For example, Katrina said that her university instructors would "guide us and scaffold us – but would not tell us the answer. It was a lot of guided questions. It was never direct teaching." Katrina "appreciated the approach" and said that she "would try to mimic it in her own teaching" – just as her father modeled it at home. Reflecting backward, she said: "It was incredibly frustrating when I was younger, but as I grew in my education, I realized what he was doing. My father remains my muse where anything science is concerned…"

Ryan Archer's Account

Ryan Archer traced his STEM and reading roots to his mother who was a full-time parent. According to Ryan, the childhood experience he most remembers is reading her chemistry book on her lap when he was three. He outlined what happened:

> Ryan: Well, it all started with my first book, my mom's chemistry book. I couldn't understand a word of it…
>
> Cheryl: And you were interested in her chemistry book?
>
> Ryan: Yes, because it was hers, and because it was old. I thought it was the coolest thing ever.
>
> Cheryl: Interesting!

Ryan: And I did not understand anything.... But I would read it. And I would read it out loud to them [my mom and dad (an engineer)]. I did not know what words were or what they were saying, but I tried. And I have always been like that [with new learning]... I don't have to understand right away, but I just jump in and love it...

The conversation continued:

Cheryl: So, you like science and mathematics?

Ryan: I had the idea that [I] would be able to do things with them, you know, like they were almost like super powers. They had that feeling for me. I thought they were cool.

In Grade 3, however, Ryan encountered a major mathematics problem:

I failed a math test...because I missed a few days, so I did not know what we were studying, and I did division, but I did it in decimal notation. I put in lines and everything like that, and when I got my grade back, I was like "How did I fail? These are all right, I am sure of it." The teacher said: "Well, you did not put the remainder on there." And I was like: "Remainder? What are you talking about?" And she said: "What do you think the 'R' is on there for?" And I said: "I don't know, I have never seen that before." Then she showed me what is leftover. I said, "Oh, you mean modalism."

He added: "I had put a little percent sign in there – and she thought I was crazy!"

This STEM story became a school trajectory-altering experience for him:

Yes, [the teacher] did not expect me to have any clue. I guess it did not hit her that what I had done was to explain through decimal notation...But my mom got that, and that is when she had a discussion with the teacher....

Ryan was assigned to gifted-and-talented classes from that point onward.

Unfortunately, gifted-and-talented education was not a good fit for him. He explained: "The programme always seemed to be...too vague. Like it lacked... structure. They wanted you to do it on your own, but they did not provide any understandable guidance."

The conversation continued:

Ryan: I needed a little bit more....

Cheryl: You needed scaffolding?

Ryan: Yes, so it was not the greatest thing to have those programmes...I'm 35 coming back to school because that is how long it took me to figure out how to direct myself – to structure my thinking...Because I have always wanted to be a teacher, but how? Without somebody building the house around me, I did not know how.

Ryan then connected his desire to be a teacher to his middle school Earth Science class where he lived in "two worlds." In his in-classroom space (Clandinin & Connelly, 1995), he was a "total nerd" with no social interactions. But outside of the classroom/school, he was the "go-to guy" that everyone came to for instruction. For years, Ryan has informally taught others like he did in middle school.

When Ryan entered university at 16, he gained "freedom," but he was "all over the place inside [his] head." According to him, he "no longer had any chains anywhere" and he "broke out all over the place." He dropped out to enter the workplace. He found satisfying employment at a technology service store. While working there, the business moved from 60th to a 10th place ranking nationally. This increase shows the kind of success Ryan found working as the employee development manager, which was a position much "like teaching." Reflecting backward, he saw this employment as part of the burgeoning teacher plotline, which had developed alongside his student plotline. In Ryan's early experiences, his tutoring of fellow students and professional development of employees helped him "grow…into the shell" of a teacher.

Upon returning to university 19 years later, Ryan fortuitously enrolled in *teach*HOUSTON and the physics as inquiry course.[3] Ryan immediately discerned the differences between physics taught as theory in high school when he was about 16 years of age, and physics taught as inquiry nearly 20 years later at university. In physics taught theoretically, he said that "the words are words dancing around in your head…that may/may not mean something to you." But, in physics as inquiry, "you come to knowledge on your own…it comes from inside of you and grows to be solid and a part of you." Ryan said that he had never had a better class in his life. The experience came at just the right time. He "jumped into" learning much like he did as a child and included his whole family in the mirror experiment, for example, where he tried to determine how large people are reflected in the mirror based on their distance from it. He shared his experience this way:

> Ryan: It was so cool to watch my mom and see her trying to be a very talented marker…
>
> Cheryl: It brought you full circle from being the three-year old reading the chemistry book to being a thirty-five-year-old coming home with a mirror?
>
> Ryan: Yes… Somehow, though, I guess I had a misconception… that it would be like being a child… Like approaching a subject… in…an innocent way…. It is not childish, though….

The conversation continued:

> Ryan: It ended up being quite the opposite. So really, you are truly like a scientist…You have to think about people like Louie McCoy who was working with microscopes and what made him so rigorous in what he was doing and to take notes the way he did…Like what led him down the path? But there you are in the process of discovery…
>
> Cheryl: Yourself, yes…
>
> Ryan: And you start to see things happen and you won't let yourself make a mistake. You start checking again…
>
> Cheryl: And you became very rigorous. Like you said, you used lots of different-sized mirrors and involved lots of people…
>
> Ryan: All of those things you learn – those things come to you [through inquiry learning].

Ryan thanked his professors and teacher educators for putting him into "the flight zone." He wants to help secondary students "get into that flight zone" and better their studies and lives without taking the circuitous path he traveled to become a science teacher.

Sam Bernard's Account

Sam's introduction to computer science was an "accident." In his view, his connections thread back to his experiences as a 10-year-old. His mother was interested in computers not only because she was a bookkeeper but also because their family was "poor" and "could not afford to buy new ones." Furthermore, his mom "loved to learn new things" and "the internet was a perfect place for her to do research…." Alongside her, Sam started "building computers." He explained:

> Mom was really interested in computers, so she always had this interesting stuff laying around…So, she taught me how to build a computer… and I have been interested in computers since then…. A bit later, I found out about "gaming" so I ended up building my own computer….

He continued:

> I wanted a [gaming] computer because my friends had great machines that could run the games…without struggling. My computer was atrocious. I would die in games… because people had not loaded onto my screen and they had already shot me! Buying a cool branded gaming machine was not feasible…so the obvious thing was to build one.

Sam not only assembled computers but he also "took other stuff apart and put it back together all of the time." The "stuff" included skateboards, rollerblades, bicycles, and the like. Sam said that whenever he experienced problems, he, like his mom, would "figure out how to fix [the] machine, frequently using the internet." He said he "gained a skill set through encountering issues and solving them on [his] own…" The next thing he knew was his friends were bringing him their computers to fix as the neighborhood "tech support."

Sam also connected his computer and fix-it talents to his father, who was technologically oriented. Of his dad, Sam said: "To this day, I don't think there is a mechanical thing he cannot figure out how to fix. He's that incredible…"

When Sam Bernard entered postsecondary education, he learned for the first time that there was a subject that had both words, *computer* and *science*. He said: "I was interested in both things and decided to pursue a computer science career." Sam, however, did not enter his current program immediately. Instead, he enrolled at a community college and "declared [his] major there." He explained:

> …And when I transferred over to the university and took my first programming class, I was hooked. [Programming] is basically a logic problem that you need to solve. It's like a logical puzzle, and I really, really enjoy doing it. So that's basically how I got into computer science…

When asked what thing is his favorite part, Sam Bernard named two courses with a professor who majors in software design. Sam said he entered the course thinking there would be "nothing to it," given his computer background. He thought it would just be another class to "go through." But then, Sam met the professor who is "barely… human [in that] he has so much knowledge to share… and the way that [the professor] teaches is very interesting…" Sam continued:

> All the knowledge he teaches is relevant to your degree… And everything he teaches is current because he works in the field and he knows what is going on… We are programming every single day, and it forces us to learn the material….

This passage later led to the following conversation:

Cheryl: So, the programming is very hands-on?

Sam: Very hands-on… He reviews your code every day, which is incredible. Even on weekends…

Cheryl: It is very inquiry-driven?

Sam: Absolutely, yeah. Because you are constantly interacting… You are in the code and you are learning it as you are writing it, and you get to an unsolvable point and you can't figure it out, so you then ask the professor and he leads you in the right direction. He won't give you the answer, he will just lead you and that's how you learn….

The exchange continued:

Cheryl: So, this sounds somewhat like how you learned to build computers with your mother?

Sam: Definitely; it is very similar. Yeah, I had to just keep on trying to figure things out… That was it. Exactly. Exactly. For me personally, it is the best way to learn.

Sam expanded on his preferred mode of learning:

Sam: You feel challenged and you don't feel frustrated. That is the best part because you can tell that you are learning because you are getting past these homework and assignment milestones. And you are getting graded, so you know you are making progress…

Cheryl: So, what have been some of the milestones in learning programming?

Sam: For me, a lot of the milestones were whenever he introduced us to a framework or a type of idea to code, like a certain mindset behind a code…It would seem like way too much information… And whenever I would begin to feel comfortable – and it may have been three weeks or may have been six weeks or may have even been the end of the course when I felt comfortable – that's when I felt I had accomplished something.

Sam's uncovering of how he learned code continued:

Cheryl: So, coming full circle was when you knew that you learned it?

Sam: Exactly.

Cheryl: And, if I were to rephrase, the information was kind of all up here and all over the place (putting her hands above her head)

Sam: Yes.

Cheryl: But when you figured it out it was…

Sam: Ingrained in me. Yeah. Absolutely.

Throughout these exchanges, how Sam learned from his professor sounded uncannily like how he learned alongside his mother and father. After his mom instructed him on how to build a computer, that knowledge became "ingrained" in Sam. He built a gaming computer on his own and assisted others with their repairs. Because of the active inquiry involving hands-on learning that Sam experienced as a child, he entered the computer science department via the community college transfer route and fruitfully learned – in an embodied way – from a professor who was so brilliant and code savvy that he was "barely human." Sam noted that what his "dad does as an air-conditioning technician is basically what [Sam] does as a programmer (troubleshoots, figures out problems, solves them)."

TALKING ACROSS THE STORIES OF EXPERIENCE

We now serially interpret the three STEM students' narratives. In our analysis, we identified the following themes that traversed the accounts: (1) familial curriculum making, (2) relationships between learners and "teachers," (3) invitations to inquiry, (4) modes of inquiry, (5) improbability of certainty, and (6) changed narratives = changed lives.

Familial Curriculum Making

Familial curriculum making occurred in Katrina's, Ryan's, and Sam's homes. Katrina was introduced to STEM through kitchen and backyard investigations with her father whereas Ryan learned about the STEM disciplines by leafing through his mother's chemistry book. As for Sam, his interest in computer science began with computer equipment his mother had scattered throughout their apartment and through participating in garage projects with his father. From the students' accounts, readers can see that "what [students] know, how [they] are bent [by their parents and others], and what [they] remember" (Schwab, 1961/1978, p. 272) congealed and brought them to this point in their beginning STEM-related careers.

Relationships between Learners and "Teachers"

Familial curriculum making, like conventional curriculum making, is dependent on healthy relationships established between learners and whomever is in the teacher position in Schwab's commonplaces. Schwab (1954/1978) declared that "if any person is to put himself…in the hands of another, he must have assurance of their gentleness in the first place and their strength and competence as well" (p. 115). In this chapter, readers see children first being and becoming learners in the hands of their parent-teacher-inquirers and how this relational family curriculum making laid the foundation for mostly productive interactions later in life with STEM professors and STEM teacher educators. In Ryan's case, we noted a considerable time lapse. A significant gap existed between him being identified as gifted-and-talented and his meeting of a caring and able teacher educator who supported his intellectual development while helping him to "grow into the shell of a [physics] teacher" from an identity perspective. We also sensed an absence of the kind of relationships Sam needed to promote learning in his narrative. Readers will recall that Sam experienced "social anxiety" during high school and did not complete his secondary education alongside his peers in the traditional manner. After-the-fact, he pursued an alternate GED pathway to gain his high school completion equivalency. Nevertheless, all three students eventually launched themselves into STEM careers, having experienced full-circle relationships between themselves as learners and a variety of "teachers."

Invitations to Inquiry

To Dewey (1910), "thinking always begins in …a forked-road situation, an [ambiguous] situation, which presents a dilemma…" (p. 11). When they were young children, Katrina, Ryan, and Sam encountered forked-road situations marked by the "suspense of uncertainty" where they wanted "to metaphorically climb tree[s]" (Dewey, 1910, p. 11). For Katrina, it was the lure of Saturday morning science school or weekend stargazing that provoked her to engage in inquiries alongside her father. For Ryan, it was his desire to read a chemistry book and understand and appreciate science in the way his mother did. And, for Sam, it was technology hardware scattered around a parent's home from which a machine analogous to a human brain could be constructed. It also was a wide array of fix-it projects in his father's workshop, which provoked him to learn. As readers can see, these parents presented their children with confounding challenges that helped them to grow academically. Whether intentional or circumstantial, the students were provided with "invitations to inquiry," as Schwab termed them. These invitations ignited their children's lifelong interests and caused them to want to solve problems currently unanswerable in their realms of experience. For Schwab, "scientific research has its origin, not in objective facts alone, but in a conception, a construction of the mind" (Schwab, 1962, p. 12). To him, "the essence of teaching science as inquiry shows some of the conclusions of science in the framework of which they arose and are tested…" (Schwab, 1962, p. 40). Tamir (1990) further noted that "the more closely [we get] to individual…projects, the more the opportunities for learning by inquiry arise" (p. 41). This certainly was the case with these three featured scholarship awardees. Katrina participated in small group learning sessions with her siblings while Sam and Ryan engaged in independent inquiry projects one-on-one with their respective parents.

Modes of Inquiry

In the students' accounts, both stable and fluid inquiries were undertaken. Stable inquiry "take[s] the principles current in the field for granted and chases the facts that the principles call for" (Schwab, 1956/1978, p. 153). This partially describes Ryan reading the chemistry book, tidbits of Katrina's kitchen physics, and elements of Sam's computer building. With a parent, each was working within the bounds of a respective discipline. For the most part, though, their parents were not delivering "rhetoric of conclusions" (Schwab, 1954/1978, p. 24) to their children. Instead, they were involving them in active learning and active testing of alternatives through informal project-based learning. This is especially true of Katrina's father. He was aware of what was not known and the need to cultivate a fluid inquiry mindset in his children. Fluid inquiry, to Schwab (1961/1978), results from "intelligent rebellion and self-education after training" (p. 242). Katrina acknowledged her father was the product of the latter when she discussed his inquiry stance. Where Sam's mother was concerned, she, too, was comfortable with alternatives because when one approach to computer assembly did not work, she simply embarked on another strategy. The same could also be said for

Sam's father whose occupation involves "troubleshoot[ing], figur[ing] problems out, [and] solv[ing] them."

The Improbability of Certainty

In Katrina's and Sam's accounts, the idea that people will not know everything all the time and the acceptance that advances in scientific fields do take place are brought to the fore. Katrina's father, through his articulated epistemological stance, directly confronted the improbability of one answer standing for all time, whereas Sam's mother, through her display of computer hardware, vicariously acknowledged that changes in technology platforms are a given. Each parent in his/her own way seemed to concur with Schwab who declared that:

> …the pervasive dynamism of things and knowledge, practical and theoretical, means that at no level of pragmatic space can education rest on inculcation only. There are no dependable patterns of reaction, no permanent catalogue of means and ends, not even a permanent body basis of all other action and reflection. (Schwab, 1959/1978, p.177–178).

Hence, all scientists, educators, and parents can do is to tell "likely stories" (Schwab, 1976, pp. 14–15) until new discoveries are made, an idea which resonates with our literature review.

While not having one answer "frustrated" Katrina as a child, she understood when she matured why her father took this position when a new planet was named in the solar system, for example. As for Sam, the changing computer parts in his family apartment gave way to his easy acceptance of different computer codes that rise in importance and fade into obscurity as new, more elegant codes replace "old ones that are no longer useful."

Changed Narratives = Changed Lives

In this chapter, readers witness how three children's narratives developed and changed as they matured into adulthood. Katrina grew into the image of an inquiry teacher that her father consistently held up as a model; Sam narrowly escaped poverty through education and finding a financially and intellectually satisfying STEM career; and Ryan altered his career trajectory by embracing a STEM curriculum making position in classrooms alongside high school youth. Also, through returning to university a second time, Ryan not only established a new career for himself but also personally learned how to scaffold his own learning through inquiry. In all three instances, the changing landscapes of the students' narratives produced changed lives, lives significantly different than where the roots of this narrative inquiry began.

FINAL WORDS

A major challenge of international science teaching is that students have "engines" – the human desire to know and natural inclinations to inquire – but often they do not have "loads" – appropriate intellectual content from the STEM fields approached through inquiry (rather than as a rhetoric of conclusions) (Schwab, 1962). Where the three scholarship students were concerned, they were active engines at young ages who encountered inquiry-driven intellectual loads they could pursue. This is because their parents accompanied them on their early academic journeys and followed where their STEM stories led. Katrina's STEM journey, as we have shown, followed the traditional pipeline route as she hit the mile markers exactly when expected, whereas Ryan and Sam arrived late at the mile markers many times over because they embarked on multiple pathway journeys to STEM. This is the missing piece that the historical and contemporary American and international literature has been unable to elaborate, despite sweeping policy claims *and* defensible claims reinforced by quantitative research only. The narrative threads we have captured became the staging grounds for the students' STEM studies as their academic trajectories show. These narratives formed critically important gateways to their current STEM careers where they now have the chance to influence the futures of thousands of others, and perhaps even their own children, as the esteemed computer science professor noted at the beginning of this chapter.

AUTHORS' NOTE

A special thank you is extended to Dr. Xiao Han and Jing Li who worked as postdoctoral and doctoral research assistants alongside Cheryl J. Craig in her role as an evaluator and/or research team member. All errors or omissions are the responsibility of the authors, not of the funding agencies.

This chapter includes findings from the following grant programmes in the United States: National Science Foundation Noyce Grant 1240083, *Recruitment, Preparation, and Retention of STEM Students as High School Teachers*, awarded to Donna W. Stokes, Paige Evans, and others; National Science Foundation Division of Undergraduate Education Grant 1356705, *Preparing Cybersecurity Students for Global Challenges of the 21st Century (S-STEM [Scholarships for STEM Education])*, awarded to Rakesh Verma and others, and the National Science Foundation Division of Graduate Education Grant 1433817, titled *Scholarship for Service: Increasing Talented Trusted Computing Professionals (SFS)*, also awarded to Rakesh Verma and others.

This chapter is reprinted with the permission of the *International Journal of Science Education*.

NOTES

1. Both grants are in no-cost extension years which hypothetically mean that new students could be awarded scholarships, although this is not unlikely.
2. The composite teaching credential would allow Katrina to teach chemistry and physics in addition to biology. The composite credential provides more flexibility in the hiring and placing of teachers. It also helps to make up for the shortage of qualified chemistry and physics teachers.
3. The NSF Noyce Grant was used in the *teach*HOUSTON STEM teacher education program.

REFERENCES

Aschbacher, P., Li, E., & Roth, E. (2010). Is science me? High school students' identities, participation and aspirations in science, engineering, and medicine. *Journal of Research in Science Teaching*, *47*(5), 564–582.

Bateson, M. C. (1994). *Peripheral visions: Learning along the way*. New York, NY: Harper Collins.

Bauer, P. J., & Fivush, R. (1992). Constructing event representations: Building on a foundation of variation and enabling relations. *Cognitive Development*, *7*(3), 381–401.

Beddoes, K. (2013). Feminist methodologies and engineering education research. *European Journal of Engineering Education*, *38*(1), 107–118.

Beijaard, D., Meijer, P., & Verloop, N. (2004). Reconsidering research on teachers' professional identity. *Teaching and Teacher Education*, *20*(2), 107–128.

Bellah, R., Madsen, R., Sullivan, W., Swidler, A., & Tipton, S. (2007). *Habits of the heart: Individualism and commitment in American life*. Berkeley, CA: University of California Press.

Blackley, S., & Howell, J. (2015). A STEM narrative: 15 years in the making. *Australian Journal of Teacher Education*, *40*(7), 101–112.

Bobbitt, F. (1924/1975). The new technique of curriculum-making. *The Elementary School Journal*, *75*, 70–77.

Bottia, M., Stearns, E., Mickelson, R., Moller, S., & Parker, A. (2015). The relationships among high school STEM learning experiences and students' intent to declare and declaration of a STEM major in college. *Teachers College Record*, *117*, 1–46.

Brown, T. (2017). The political shaping of teacher education in STEM areas. In J. Clandinin & J. Husu (Eds.), *SAGE handbook of research in teacher education* (pp. 665–680). Thousand Oaks, CA: SAGE Publications.

Bruner, J. (1987). Life as narrative. *Social Research*, *54*, 11–32.

Bruner, J. (2002). *Making stories: Law, literature and life*. New York, NY: Farrar, Strauss and Giroux.

Bruner, J. (2010). Narrative, culture, and mind. In T. A. Nylund, A. De Fina, & D. Schiffrin (Eds.), *Telling stories: Language, narrative, and social life* (pp. 45–49). Washington, DC: Georgetown University Press.

Bybee, R. (2013). *The case for STEM education: Challenges and opportunities*. Arlington, VA: National Science Teachers Association.

Cannady, M., Greenwald, E., & Harris, K. N. (2014). Problematizing the STEM pipeline metaphor: Is the STEM pipeline metaphor serving our students and the STEM workforce? *Science Education*, *98*(3), 443–460.

Carlone, H., & Johnson, A. (2007). Understanding the science experiences of successful women of color: Science identity as an analytic lens. *Journal of Research in Science Teaching*, *44*(8), 1187–1218.

Case, J., & Light, G. (2011). Emerging research methodologies in engineering education research. *Journal of Engineering Education*, *100*(1), 186–210.

Chang, M., Eagan, M., Lin, M., & Hurtado, S. (2011). Considering the impact of racial stigmas and science identity: Persistence among biomedical and behavioral science aspirants. *The Journal of Higher Education*, *82*(5), 564–596.

Clandinin, D. J., & Connelly, F. M. (1995). *Teachers' professional knowledge landscapes*. New York, NY: Teachers College Press.

Clandinin, D. J., & Connelly, F. M. (2000). *Narrative inquiry: Experience and story in qualitative research*. San Francisco, CA: Jossey-Bass.

Cleaves, A. (2005). The formation of science choices in secondary school. *International Journal of Science Education*, *27*(4), 471–486.

Coleman, J., Campbell, E., Hobson, C., McPartland, J., Mood, A., Weinfeld, F., & York, R. (1966). *Equality of educational opportunity*. Washington, DC: U.S. Government Printing Office.

Conle, C. (1996). Resonance in preservice teacher inquiry. *American Educational Research Journal*, *33*(2), 297–325.

Conle, C. (2003). An anatomy of narrative curricula. *Educational Researcher*, *32*(3), 3–15.

Connelly, F. M., & Clandinin, D. J. (1990). Stories of experience and narrative inquiry. *Educational Researcher*, *19*(5), 2–14.

Connelly, F. M., & Clandinin, D. J. (1999). *Shaping a professional identity: Stories of educational practice*. New York, NY: Teachers College Press.

Connelly, F. M., & Clandinin, D. J. (2005). Narrative inquiry. In J. Green, G. Camilli, & P. Elmore (Eds.), *Complementary methods for research in education* (3rd ed., pp. 477–488). Washington, DC: American Educational Research Association.

Craig, C., Evans, P., Stokes, D., Bott, S., & Abrol, B. (2017a). Attracting, preparing and retaining teachers in high need areas: A science as inquiry model of teacher education. In M. Peters, B. Cowie, & I. Menter (Eds.), *A companion for teacher education research*. Dordrecht: Springer Publications.

Craig, C., Zou, Y., & Curtis, G. (2017b). The developing knowledge and identity of an Asian-American teacher: A narrative inquiry into the influence of a China study abroad experience. *Learning, Culture, and Social Interaction*, *17*, 1–20.

Craig, C. (2007). Story constellations: A narrative approach to contextualizing teachers' knowledge of school reform. *Teaching and Teacher Education*, *23*(2), 173–188.

Craig, C. (2011). Narrative inquiry in teaching and teacher education. In J. Kitchen, D. Ciuffetelli Parker, & D. Pushor (Eds.), *Narrative inquiries into curriculum making in teacher education* (pp. 19–42). Bingley: Emerald Publishing Limited.

Crawford, B. (2000). Embracing the essence of inquiry: New roles for science teachers. *Journal of Research in Science Teaching*, *37*(9), 916–937.

Derry, S., Pea, R., Barron, B., Engle, R., Erickson, F., Goldman, R., & Sherin, B. (2010). Conducting video research in the learning sciences: Guidance on selection, analysis, technology, and ethics. *The Journal of the Learning Sciences*, *19*(1), 3–53.

Dewey, J. (1910). *How we think*. London: DC and Heath and Company.

Dewey, J. (1916). *Education and democracy*. New York, NY: Macmillan Press.

Dewey, J. (1938). *Experience and education*. New York, NY: Basic Books.

Drake, C., & Sherin, M. (2006). Practicing change: Curriculum adaptation and teacher narrative in the context of mathematics education reform. *Curriculum Inquiry*, *36*(2), 153–187.

Driver, R., Asoko, H., Leach, J., Scott, P., & Mortimer, E. (1994). Constructing scientific knowledge in the classroom. *Educational Researcher*, *23*(7), 5–12.

Duit, R., & Treagust, D. (2003). Conceptual change: A powerful framework for improving science teaching and learning. *International Journal of Science Education*, *25*(6), 671–688.

Egan, K. (1985). Teaching as story-telling: A non-mechanistic approach to planning teaching. *Journal of Curriculum Studies*, *17*(4), 397–406.

Erickson, F. (2012). Qualitative research methods for science education. In B. Fraser, et al. (Eds.), *Springer international handbook of education* (pp. 1451–1469). Dordrecht: Springer.

Fivush, R., & Buckner, J. (2003). Creating gender and identity through autobiographical narratives. In R. Fivush & C. Haden (Eds.), *Autobiographical memory and the construction of a narrative self: Development and cultural perspectives* (pp. 149–167). Mahwah, NJ: Lawrence Erlbaum Associates Publishers.

Fivush, R., & Nelson, K. (2004). Culture and language in the emergence of autobiographical memory. *Psychological Science*, *15*, 573–577.

Fleischer, H. (2012). What is our current understanding of one-to-one computer projects? A systematic narrative research review. *Educational Research Review*, *7*(2), 107–122.

Freeman, B., Marginson, S., & Tytler, R. (Eds.). (2014). *The age of STEM: Educational policy and practice across the world in science, technology, engineering and mathematics*. New York, NY: Routledge.

Gardner, D. (1983). *A nation at risk*. Washington, DC: The National Commission on Excellence in Education, US Department of Education.

Goodson, I. (1990). Studying curriculum: Toward a social constructionist perspective. *Journal of Curriculum Studies*, *22*(4), 299–312.

Hanushek, E. (2016). What matters for student achievement: Updating Coleman on the influence of families and schools. *Education Next*, *16*(2). Retrieved from http://educationnext.org/what-matters-for-student-achievement/

Hazari, Z., Sadler, P., & Sonnert, G. (2013). The science identity of college students: Exploring the intersection of gender, race, and ethnicity. *Journal of College Science Teaching*, *42*(5), 82–91.

Henricksen, E., Dillon, J., & Ryder, J. (2015). *Understanding student participation and choice in science and technology education*. Singapore: Springer Nature.

Hollingsworth, S., Dybdahl, M., & Minarik, L. T. (1993). By chart and chance and passion: The importance of relational knowing in learning to teach. *Curriculum Inquiry*, *23*(1), 5–35.

Huber, J., Murphy, M. S., & Clandinin, D. J. (2011). *Places of curriculum making: Children's lives in motion*. Bingley: Emerald Publishing Limited.

Hwang, S. (2011). Narrative inquiry for science education: Teachers' repertoire-making in the case of environmental curriculum. *International Journal of Science Education*, *33*(6), 797–816.

Jackson, P., & Seiler, G. (2013). Science identity trajectories of latecomers to science in college. *Journal of Research in Science Teaching*, *50*, 826–857.

Jencks, C., Smith, M., Acland, H., Bane, M. J., Cohen, D., Gintis, H., ... Michelson, S. (1972). *Inequality: A reassessment of the effect of family and schooling in America*. New York, NY: Basic Books.

Kaasila, R. (2007). Using narrative inquiry for investigating the becoming of a mathematics teacher. *Zentralblatt für Didaktiks der Mathematik (ZDM)*, *39*(3), 205–213.

King, T. (2003). *The truth about stories: A native narrative*. Toronto, CA: House of Anansi Press, Inc.

Kuhn, T. (1970/1996). *The structure of the scientific revolution* (3rd ed.). Chicago, IL: University of Chicago Press.

Lerman, S. (2000). The social turn in mathematics education research. In J. Boaler (Ed.), *Multiple perspectives on mathematics teaching and learning* (pp. 19–44). Westport, CT: Ablex Publishing.

Luhman, J. T., & Boje, D. M. (2001). What is complexity science? A possible answer from narrative research. *Emergence: A Journal of Complexity Issues in Organizations and Management*, *3*(1), 158–168.

Lyons, N., & LaBoskey, V. (2002). *Narrative inquiry in practice: Advancing the knowledge of teaching*. New York, NY: Teachers College Press.

Maxwell, R. (2001). Why culture works. In *Culture works: The political economy of culture* (pp. 1–22). Minneapolis, MN: University of Minnesota Press.

Mishler, E. (1990). Validation in inquiry-guided research: The role of exemplars in narrative studies. *Harvard Educational Review*, *60*(4), 415–443.

Nash, R. (2014a). If stories heal, then what exactly is a story? In R. Nash & S. Viray (Eds.), *How stories heal* (pp. 34–43). New York, NY: Peter Lang.

Nash, R. (2014b). Writing our way into healing, meaning and wholeness: A personal example. In R. Nash & S. Viray (Eds.), *How stories heal* (pp. 16–33). New York, NY: Peter Lang.

National Academy of Science. (2010). *Rising above the gathering storm, revisited: Rapidly approaching category 5*. Washington, DC: National Academies Press.

National Research Council. (2013). *Monitoring progress toward successful K-12 STEM education: A nation advancing?* Washington, DC: National Academies Press.

National Research Council. (2014). *STEM integration in K-12 education: Status, prospects, and an agenda for research*. Washington, DC: National Academies Press.

Olson, M. (2000a). Curriculum as a multistoried process. *Canadian Journal of Education/Revue canadienne de l'éducation*, *25*(3), 169–187.

Olson, M. (2000b). Where the story leads: A response to Cheryl Craig. *Canadian Journal of Education/ Revue canadienne de l'éducation*, *25*(4), 349–352.

President's Council of Advisors on Science and Technology (U.S.). (2010). *Prepare and inspire: K-12 education in science, technology, engineering and mathematics (STEM) for America's future: Executive report*. Washington, DC: Executive Council of the President, President's Council of Advisors on Science and Technology, President's Council of Advisors on Science and Technology.

Pushor, D. (2013). Bringing into being a curriculum of parents. In D. Pushor & The Parent Engagement Collaborative (Eds.), *Portals of promise: Transforming beliefs and practices through a curriculum of parents* (pp. 5–19). Rotterdam, NL: Sense Publishers.

Ratelle, C., Larose, S., Guay, F., & Senecal, C. (2005). Perceptions of parental involvement and support as predictors of college students' persistence in a science curriculum. *Journal of Family Psychology*, *19*(2), 286–293.

Rozek, C., Svoboda, R., Harackiewicz, J., Hulleman, C., & Hyde, J. (2017). Utility-value intervention with parents increases students' STEM preparation and career pursuit. *Proceedings of the National Academy of Sciences*, *114*(5), 909–914.

Sack, J., Quander, J., Redl, T., & Leveille, N. (2016). The community of practice among mathematics and mathematics education faculty members at an urban minority serving institution in the US. *Innovative Higher Education*, *41*(2), 167–182.

Schwab, J. J. (1954/1978). Eros and education: A discussion of one aspect of discussion. In I. Westbury & N. Wilkof (Eds.), *Science, curriculum and liberal education: Selected essays*. Chicago, IL: University of Chicago Press.

Schwab, J. J. (1956/1978). Science and civil discourse: The uses of diversity. In I. Westbury & N. Wilkof (Eds.), *Science, curriculum and liberal education: Selected essays*. Chicago, IL: University of Chicago Press.

Schwab, J. J. (1959/1978). The 'impossible' role of the teacher in progressive education. In I. Westbury & N. Wilkof (Eds.), *Science, curriculum and liberal education: Selected essays*. Chicago, IL: University of Chicago Press.

Schwab, J. J. (1961/1978). Education and the structure of the disciplines. In I. Westbury & N. Wilkoff (Eds.), *Science, curriculum and liberal education: Selected essays*. Chicago, IL: University of Chicago Press.

Schwab, J. J. (1962). The teaching of science as enquiry. In J. Schwab & P. Brandwein (Eds.), *The teaching of science* (pp. 3–103). Cambridge, MA: Harvard University Press.

Schwab, J. J. (1969). The practical: A language for curriculum. *The School Review*, *78*(1), 1–23.

Schwab, J. (1973). The practical 3: Translation into curriculum. *The School Review*, *81*(4), 501–522.

Schwab, J. J. (1976). *Genesis: A discussion of principles of inquiry*[transcript of cassette recording]. East Lansing, MI: Michigan State University.

Schwartz, K. (2015, June 5). Could storytelling be the secret sauce to STEM education? KQED public media for Northern California. Retrieved from https://ww2.kqed.org/mindshift/2015/06/05/could-storytelling-be-the-secret-sauce-to-stem-education/

Spence, D. (1984). *Narrative truth and historical truth: Meaning and interpretation in psychoanalysis*. New York, NY: Norton.

Stone, E. (1988). *Black sheep and kissing cousins: How our family stories shape us*. New York, NY: Penguin Books.

Stuckey, M., Hofstein, A., Mamlok-Naaman, R., & Eilks, I. (2013). The meaning of 'relevance' in science education and its implications for the science curriculum. *Studies in Science Education*, *49*(1), 1–34.

Tamir, P. (1990). Considering the role of invitation in science teaching and in teacher education. *Journal of Science Teacher Education*, *1*(3), 41–45.

Taylor, P., & Wallace, J. (2007). Contemporary qualitative research for science and mathematics educators. In P. Taylor & J. Wallace (Eds.), *Qualitative research in post-modern times* (pp. 1–11). Dordrecht: Springer.

Van Driel, J., & Berry, A. (2012). Teacher professional development focusing on pedagogical content knowledge. *Educational Researcher*, *41*(1), 26–28.

Van Driel, J. H., Beijaard, D., & Verloop, N. (2001). Professional development and reform in science education: The role of teachers' practical knowledge. *Journal of Research in Science Teaching*, *38*(2), 137–158.

Viray, S. (2014). Final letter to our readers. In R. Nash & S. Viray (Eds.), *How stories heal* (pp. 187–191). New York, NY: Peter Lang.

Wallace, C. S. (2012). Authoritarian science curriculum standards as barriers to teaching and learning: An interpretation of personal experience. *Science Education*, *96*(2), 291–310.

Xu, S., & Connelly, F. M. (2009). Narrative inquiry for teacher education and development: Focus on English as a foreign language in China. *Teaching and Teacher Education*, *25*(2), 219–227.

IN PRAISE OF "UNSUNG TEACHERS": TEACHERS' INFLUENCES ON STUDENTS ENROLLING IN STEM PROGRAMS WITH THE INTENT OF ENTERING STEM CAREERS

Cheryl J. Craig, Paige K. Evans, Rakesh Verma, Donna W. Stokes and Jing Li

ABSTRACT

This narrative inquiry examines teachers' influences on undergraduate/graduate students who enrolled in STEM programs and intended to enter STEM careers. Three National Science Foundation (NSF) scholarship grants sat in the backdrop. Narrative exemplars were crafted using the interpretative tools of broadening, burrowing, storying and restorying, fictionalization, and serial interpretation. Three diverse students' narratives constituted the science education cases: one from teacher education, another about cybertechnology, and a third involving cybersecurity. The influence of the university students' former teachers cohered around five themes: (1) same program-different narratives, (2) in loco parentis, (3) counterstories, (4) learning in small moments, and (5) the importance of the liberal arts in STEM education. The students' narratives form instructive models for their siblings and other students pursuing STEM degrees/careers. Most importantly, the multi-perspectival stories of experiences capture the far-reaching impact of "unsung teachers" whose long-term influence is greatly underestimated by the public.

Preparing Teachers to Teach the STEM Disciplines in America's Urban Schools
Advances in Research on Teaching, Volume 35, 135–158

ISSN: 1479-3687/doi:10.1108/S1479-368720210000035008

Keywords: Teacher education; teacher influence; teachers' long-term impact; narrative inquiry; STEM careers; *teach*HOUSTON

> An army of unsung teachers helped me along the way. That's what teachers do, you know… –Student Participant in NSF Division of Undergraduate Education Noyce Grant Program (*teach*HOUSTON)

> I can't remember their names and faces… But I do know teachers put me back on my feet whenever I floundered… –Student Participant in NSF Division of Graduate Education Cybercorps Grant Program

> Some provided meals and snacks when my hunger pangs were too intense for me to learn; others gave moral support and hope for the future. –Student Participant in NSF Division of Undergraduate Education S-STEM Grant Program

> Seventeen undergraduate students are studying the STEM disciplines at [this research-intensive university] because of an Advanced Placement Calculus teacher at a high school near the ship channel… We still meet all the time… We are like family… –Student Participant in NSF Division of Undergraduate Education S-STEM Grant Program

When we traversed the field texts gathered from three US National Science Foundation (NSF) grant initiatives providing scholarships to Science | Technology | Engineering | Mathematics (STEM) undergraduate and graduate students interested in entering discipline-related careers (one STEM teacher education grant [*teach*HOUSTON], one undergraduate cybertechnology grant, one graduate cybersecurity grant), several research themes from which knowledge could be disseminated locally, nationally, and internationally rose to the fore. Among the topics were the value of scholarship programs to young people's potential careers and lives and the influence of parents, professors, and teachers on students' entering STEM fields and STEM-related occupations. In this chapter, we focus on the influence of teachers on primary and secondary students who subsequently enrolled in university-based STEM programs with the intention of entering discipline-related professions. For readers' information, this chapter on the impact of parents (Craig, Verma, Stokes, Evans, & Abrol, 2018) has already been published.

From the outset, this work on the influence of teachers took a somewhat different path to the impact of parents and professors papers. Where the said chapter was concerned, stories about named parents and professors bubbled to the surface in the individual and combined data pools. However, when teachers were mentioned, they were discussed in narrative fragments like the excerpts we used to introduce our research phenomenon and to open our paper. Put differently, the influence of teachers was communicated in small stories rather than in large narratives. Furthermore, the teachers in question were not identified by name nearly as often as were the professors and parents. The undergraduate and graduate students' primary and secondary educators tended to be remembered as "unsung teachers…[doing] what teachers do" – to borrow the apt turn of phrase used by a *teach*HOUSTON student whose way with words was used to launch this chapter and to form its title.

We begin this work with our literature review, which is followed by our conceptual framework. After that, we outline our research methodology where we pay close attention to the role of small stories in narrative research. Next, we share three exemplars that instantiate the influence of teachers on undergraduate and graduate students' enrolling in the STEM disciplines and subsequently entering STEM-related careers. These elucidations breathe life into the small stories nested within metanarratives that were lived and relived, and told and retold, by the two undergraduate students and one graduate student who constituted our sample of convenience for this research investigation. To conclude, we analyze and synthesize what can be learned about the short-term and long-term impact of teachers on students' tertiary education and career choices.

LITERATURE REVIEW

Our survey of the literature revolves around two strands of research: (1) the influence of teachers on student success and (2) the use of narrative and narrative research in STEM education.

Influence of Teachers on Student Success

In the United States, the Coleman Report (Coleman et al., 1966), along with the Jencks Report (Jencks et al., 1972) and other contemporary American literature, has consistently upheld that teachers impact student success, especially where minority youth are concerned. The Coleman Report is known for being the most significant educational document of the twentieth century as it set the American educational policy backdrop during the civil rights era and beyond. The survey research report asserted that quality of teachers shows a stronger relationship [than school facilities and curricula] to pupil achievement [particularly for students of color]. Furthermore, it is progressively greater at higher grades, indicating a cumulative impact of the qualities of teachers…. on pupils' achievements (Coleman et al., 1966, as cited in Goldhaber, 2016). The finding that "teacher quality is one of the few school characteristics that significantly affects student performance" is an influence that has continued to reverberate in the national and international literature, for example, in an OECD (2005) report having to do with *Teachers matter*, UNICEF's (2009) Child-Friendly Schools Program and UNESCO's (2011) teacher competencies framework.

Nationally, Linda Darling Hammond reviewed the policy documents from 50 American states to determine the relationship between teacher quality and student achievement. She concluded that "the findings of both the qualitative and quantitative analyses suggest that policy investments in the quality of teachers may be related to improvements in student performance" (Darling-Hammond, 2000, p. 1). Furthermore, "the effects of well-prepared teachers on student achievement [appear to] be stronger than the influences of student background factors, such as poverty, language background, and minority status" (p. 33). Additionally, quantitative value-added assessment studies of teacher effects in the classroom have also found that differential teacher effectiveness contributes to differences in student learning more than the effects of differences in class size and heterogeneity (Jordan, Mendro, & Weersinghe, 1997; Sanders & Rivers, 1996; Wright, Horn, & Sanders, 1997). Furthermore, students consecutively assigned ineffective teachers have significantly lower achievement and lower gains in achievement than those assigned to several highly effective teachers in sequence (Sanders & Rivers, 1996). Based on her own research and informed by the value-added findings, Darling-Hammond (2000) concluded that teacher effects are "additive and cumulative, and generally not compensatory" (p. 2). This is consistent with the results reported by the National Commission on Teaching and America's Future (1996) and with Goldhamer's (2016) more recent article in *EducationNEXT*, in addition to the policies and reports of the international agencies cited earlier.

Two other meta-analyses of the international literature on teacher influences also point to the significant impact of teachers on student achievement and attainment and bear mentioning. One was undertaken by John Hattie; the other was conducted by Debora Roorda and her research team members. Hattie (2008) focused his study on evidence supporting teachers making a difference on student achievement. He synthesized over 800 meta-analyses related to student achievement. Hattie determined that students themselves are responsible for 50% of the variance in their achievement test scores, that their homes, schools, principals, and peers were each responsible for 5%–10% of their variance in scores, and that their teachers made up 30% of the variance in student test results (Hattie, 2003). According to Hattie (2003, 2008), the feedback, instructional quality, styles, and questioning strategies of teachers, among other considerations, contributed to their effects on students. Hattie also made fine-line distinctions between expert teachers and experienced teachers with expert teachers having the most effect. As for Roorda, Koomen, Split, and Oort (2011), they more recently conducted a meta-analysis of 99 studies. Their overarching research question had to do with the influence of affective teacher–student relationships on students' school engagement and achievement. They found that engagement teacher effects were medium to large, while associations with student achievement were small to medium. Interestingly, Roorda, Koomen, Spilt, and Oort (2011) found stronger effects in the higher grades while negative effects were most profound in the lower grades.

Having established through mostly national and international quantitative research studies that teachers significantly impact students' experiences, engagement, and achievement, we now focus attention on the second strand of this literature review: narrative research in STEM education.

Narrative Research in STEM Education

Narrative inquiry has experienced a "meteoric rise" (Lyons, 2010, p. 600) in nearly every disciplinary field, including computational mathematics. In fact, the influence of narrative has become so pervasive that Schwartz (2015) questioned whether it may be the "secret sauce" of which STEM education has not fully taken advantage.

A comprehensive literature review about the use of narrative inquiry as a qualitative research method in the fields of teaching and teacher education has already been conducted. That review (Craig, 2011) pinpointed STEM-related narrative inquiries in biology (Chang & Rosiek, 2003), elementary mathematics (Ross, 2003, 2004), and secondary mathematics (Sack, 2008). Since 2011, narrative research has been used in tertiary education by Mathematics and Mathematics Education faculty at a minority serving university (Sack, Quander, Redl, & Leveille, 2016) and in a mathematics, chemistry, and physics teacher education program (Craig, Evans, Bott, Stokes, & Abrol, 2017). Further examples of narrative employed in the STEM disciplines can be found in the *Journal of Science Teacher Education* (i.e., Dietz & Davis, 2009; Wagler, 2010) and the *Journal of Research for Science Teaching* (i.e., Kurth, Kidd, Gardner, & Smith, 2002; Varelas, Kane, & Wyle, 2012). Cumulatively, these articles show the symbiotic relationship that exists between the narrative inquiry research method where inquiring into narratives forms the *modus operandi*, and the subject matter of the STEM disciplines being both taught and learned through the scientific inquiry process. The "secret sauce" to which Schwartz earlier referred might very well be the tight connections historically existing between and among STEM, narrative and inquiry. As an aside, it was no coincidence that Connelly and Clandinin (1986) published their inaugural narrative inquiry articles in the *Journal of Research for Science Teaching*. Connelly's subject area specialty was science (biology); his doctoral supervisor was Joseph Schwab (1960). Undoubtedly, Schwab's research on scientific inquiry contributed to Connelly and Clandinin's (1990) groundbreaking work on narrative inquiry. These science-inquiry associations will become even more evident as this chapter unfurls.

CONCEPTUAL FRAMEWORK

Four key concepts underpinning narrative inquiry research frame this study: experience, story, identity, and curriculum making. Each conceptualization will be discussed with respect to the influence of teachers on primary and secondary students who subsequently enrolled in university-based STEM programs in the hope of entering STEM-related professions.

Experience

Experience is "a changing stream…characterized by continuous interaction of human thought with [the] personal, social and material environment" (Clandinin & Rosiek, 2006, p. 39). At the intersection where this shifting stream of life begins, experience and education start as well. This is because life fuels both experience and education (Dewey, 1938). In this research study, students' experiences form an integral part of their lives from birth to high school to university and beyond. According to Dewey (1938), experience has both active and passive elements to it, with the active part being the students' attempts at "trying" and the passive part being their simultaneous "undergoing." In other words, when students experience something, they act upon it while simultaneously having to live with its effects (Dewey, 1916, p. 163). Furthermore, current experiences build on past experiences to inform future experiences (Dewey, 1938). On this past–present–future continuum, experience cumulatively shapes how students live, how they learn, and ultimately, who they are. This point is important to keep in mind because what we focus on in this chapter are students' retrospective recollections of their teachers preparing them for the STEM disciplines as youngsters. The expansion of the students' streams of experience and their reflections on those experiences over time with multiple teachers also come into play.

Story

Experience is also intimately connected to narrative. In fact, experience can only be made visible through storytelling and represented in narrative form. Reflection is similarly implicated because storying and restorying involves probing changes and continuities through looking backward and forward, inside and out (Clandinin & Connelly, 2000). "Following where [stories] lead" (Olson, 2000) is how the contours of students' educational trajectories take shape.

According to Bruner (2002), "stories start early and continue ceaselessly" (p. 1) throughout students' lives. Through the exchange of stories, "a child's education is instantly, naturally, and authentically underway" (Pushor, 2013, p. 7). This means that youth grow up with certain narratives "under their skin." Some tales are "weightless pleasures"; others are negatively imprinted on their psyches (Stone, 1988, p. 10). Narratives "model not only a world, but minds seeking to give [the world] meaning" (Bruner, 2010, p. 27). Through the sharing of personal narratives, the world as embodied in the stories students live and tell becomes known.

Identity

In narrative research, the term, "stories to live by" (Connelly & Clandinin, 1999) connotes identity. The concept of "stories to live by" relates to how "our identities [are] composed and improvised as we go about living our lives embodying knowledge and engaging our contexts" (Connelly & Clandinin, 1999, p. 4). "Stories to live by" are both personal (reflecting students' school and life histories) and social (reflecting their milieus) including the formal and informal learning milieus, in which students have interacted. In other words, students' "stories to live by" – their identity-related narratives – are shaped by their knowledge forged in multiple contexts. Concurrently, students' "stories to live by" are affected by the different landscapes in which they have lived and the kinds of feedback they have received from others, including teachers (Clandinin & Connelly, 1995). The feedback they experience "give[s] messages and instructions; ...offer[s] blueprints and ideals; [and]... issue[s] warnings and prohibitions" (Stone, 1988, p. 5), which, in turn, influences their identities. In this way, students' identities are

> ...an evolving nexus where all the [streams of experience] that constitute life converge in the ... self: genetic makeup, the nature of the man and woman who gave life, the culture in which [students are] raised, people who have sustained [students] and people who have done [them] harm, the good and ill [they] have done to others and to [themselves], the experience of love and suffering – and much, much more. In the midst of this complex field, identity is a moving intersection of the inner and outer forces that make [youths] who [they are].... (Palmer, 1998, p. 13)

Within this complex constellation (Craig, 2007) of shaping forces, individual teachers play an important, though transitory, role. For one or more academic years, they spend more time with students daily than their parents do. As Schwab (1983) put it, students "are better known by no one [else] but the teacher" because the teacher is the only one who actively "tries to teach them"; he/she is the only one "who lives with them for the better part of the day and the better part of the year" (p. 245). Because of the intensity of the teaching–learning relationship, teachers are positioned to shape what students come to know, what they do, and who they are. Individual youths, however, retain the power to dramatically change their "stories to live by" and alter the course of their unfolding lives (King, 2003) at any juncture of time.

Curriculum Making

Every curriculum making situation requires a teacher, a learner, subject matter, and milieu working in sync with one another and contributing in equal ways (Schwab, 1973). These four irrefutable curriculum commonplaces are essential for curriculum making to take place.

Teacher

For Schwab, teachers – the first commonplace to be described – are "agents of education, not [of] its subject matter" (1954/1978, p. 128). Schwab made this distinction in the passage below:

> [The teacher] wants something more for his students than the capacity to give back to him a report of what he himself has said. He wants them to possess a knowledge or a skill in the same way that he possesses it, as a part of his best-loved self...He wants to communicate some of the fire he feels, some of the Eros he possesses, for a valued object. His controlled and conscious purpose is to liberate, not captivate the student. (Schwab, 1954/1978, pp. 124–125)

In other words, Schwab saw teachers as opening up worlds and possibilities for careers for students rather than chaining them to the subject matter knowledge they (the teachers) teach.

Learner

The second curriculum commonplace is the learner. Learners enter the teaching–learning situation with "different personal histories which confer on

[them] widely varying wants and capacities for satisfaction" (Schwab, 1959/1978). Schwab further emphasized that "what [learners] are, what [learners] know, how [learners] have been bent, and what [learners] remember, determine what [learners] experience..." (Schwab, 1961/1978, p. 272) in the curriculum making act.

Subject Matter

Where the third commonplace, subject matter, is concerned, Schwab wrote that the disciplines have no fixed catalog of structure (Schwab, 1961/1978). To him, *nobody* knows the structure of the disciplines (p. 243; italics in original) because there are no "natural joints" of discipline that occur in nature (Schwab, 1980, p. 365). Because Schwab rebelled against the certainty – the rhetoric of conclusions – that stable inquirers attribute to science (including education understood as a scientific enterprise), he believed that "facts" in textbooks and other documents should be presented as "tentative formulations – not facts, but *interpretations* of facts" (p. 242; italics in original). Thus, subject matter in lessons and textbooks, to his way of thinking, would be filled with "examples of the uncertainties, the differences of interpretation, and the issues of principle which characterize the disciplines."

Milieu

Schwab's linking of subject matter with inquiry also sheds light on his understanding of the fourth curriculum commonplace, milieu. For Schwab, "knowledge of a context of discourses [Note: multiple discourses; not one discourse] gives us...a fuller knowledge of the scope and meaning of the conclusions...." This led Schwab to declare that "To the question, how big a context? There is no clear answer. There is yet more to know or more to know about..." (Schwab, 1956/1978, p. 153). Here the impulse to inquire coupled with the indeterminable nature of context distinguishes Schwab's rendering of milieu. Also, the phrases, to know and to know about, imply that, for Schwab, there would be infinitely more to do and think about doing. This disciplined approach to knowing and doing eventually led to Schwab's outright declaration that "the problems of education arise from exceedingly complex actions, reactions, and transactions... these doings constitute a skein of myriad threads which know no boundaries..." (Schwab, 1971/1978, p. 329).

RESEARCH METHODOLOGY

Introduction to Narrative Inquiry

In narrative inquiry research, narrative serves as both the method of investigation and the phenomenon of inquiry (Connelly & Clandinin, 1986; Connelly & Clandinin, 1990). Reduced to its essence, narrative is "the portal through which a person enters the world" (Connelly & Clandinin, 2005, p. 477). Simply stated, story is the "portal to experience" (Xu & Connelly, 2010, p. 352). Because of this, narrative inquirers intentionally avoid strategies, rules, and techniques (Clandinin & Connelly, 2000) from other places because "life is not made up of separate pieces" (Bateson, 1994, p. 108). Teaching and learning are personal and emotional as well as cognitive and rational (Hollingsworth, Dybdayl, & Minarik, 1993). Narrative inquiry is a relational form of research with three commonplaces: temporality, sociality, and place/situation (Dewey, 1938). Clandinin and Connelly's use of the term, commonplace, directly connects the method's roots to Schwab's (1969) "practical, a language for curriculum" and his commonplaces of curriculum, whereas temporality, sociality, and place/situation locate the origins of the narrative inquiry research method in Dewey's (1938) philosophy of experience. Hence, for a narrative inquiry to reflect its field-based lineage (Xu & Connelly, 2010), all three commonplaces must be present and interacting simultaneously.

Sources of Evidence

Three forms of evidence constitute the storied data pool for this chapter. The first two are regularly scheduled interviews and focus group sessions with the graduate and undergraduate students and their professors, which occurred for all three funded grant projects. The third piece of evidence is reflective writing in journals or via email communications as some students graduated and entered the workforce in other parts of the country.

Tools of Analysis

Three analytical tools – broadening, burrowing, and storying and restorying (Connelly & Clandinin, 1990) – are used to report the storied evidence in this research study. A fourth tool – fictionalization – was added to the three initial tools to further protect the privacy of students' personal narratives and to allow complex, contested narratives to be told and retold while they continue to be lived and relived.

Broadening

Through broadening, we historically trace our work to the Coleman Report, the first American policy document which asserted that teacher quality affects student outcomes, particularly where students of color are concerned. We similarly use broadening to show the tight bond that has been forged between scientific inquiry (Schwab) and narrative inquiry (Connelly and Clandinin) as well as the strong link between Dewey's theory of experience and Connelly and Clandinin's narrative inquiry research method.

Burrowing

By engaging in burrowing, our second research tool, we dig deeply into particular aspects of the stories of three undergraduate and graduate students in our sample of convenience. We particularly unpack – burrow into – the influence of teachers on the educational trajectories of undergraduate and graduate students studying the STEM disciplines with the intent of entering STEM-related careers. Through burrowing, we illuminate teachers' impact on primary and secondary students and how that influence continues to unfurl in the students' tertiary education environment and beyond.

Storying and Restorying

The third analytical tool, storying and restorying, allows us to animate instances of prior curriculum making and how those past situations connect to the three students' current participation in the scholarship grant programs and engage in possible/current STEM careers. Storying and restorying becomes evident when narrative inquirers skillfully interweave texts from various other data sources (multiple interviews, focus group participation, email/cell phone communication).

Fictionalization

Fictionalization (Caine et al., 2016; Clandinin, Murphy, Huber, & Orr, 2009) is the fourth interpretive device we use in this narrative inquiry. We use fictionalization to shift situations in ways that mask students' identities, so they are not as recognizable to others, especially professors and classmates immediately around them. Our intent is to respect students' privacy to the degree possible while telling experientially rich stories of their lives. This means we purposely leave out details that are nonessential to this chapter's storyline. Through shifting some recognizable narrative material, we tell stories previously not made public because of their sensitive natures.

Serial Interpretation

In addition to the four analytical devices we have sketched, this narrative inquiry also engages in overarching, cross-case analysis called "serial interpretation" (Schwab, 1954/1978). Serial interpretation increases explanatory power and gives the findings of multiple studies added heft. It serves as a vehicle to discuss "encompassing idea[s]" that transcend multiple research investigations. Serial interpretation involves "talking across" (Stone, 1988, p. 2) or "see[ing] across" (Clandinin, 2013, p. 131) several research projects, rather than reporting a single case or a single grant's results. It recently has been used by Craig (2018) to unpack metaphors of knowing, doing, and being in teaching and teacher education arising from multiple research studies and by Craig, You, et al. (2718) to analyze the nature of embodied knowledge in cross-case, funded projects. These latter two articles form vital pieces of the foundation on which this research study is built.

"SMALL STORIES" AND MEGANARRATIVES

Following in the footsteps of Olson and Craig (2009) and Ciuffetelli Parker and Craig (2017), this research also spotlights "atypical, noncanonical 'small stories'" (Georgakopoulou, 2004) of curriculum making that took place between teachers and students. In this chapter, formal and informal curriculum making between three current undergraduate and graduate students and their former elementary and secondary teachers is featured. These personalized, contextualized small stories are presented in juxtaposition to STEM education's meganarratives. Meganarratives or "grand stories," as Cohen and Garet (1975) called them, cohere around "large and loose set[s] of ideas about how [STEM education] works, why it goes wrong and how it can be set right" (p. 21). Grand stories are very influential because educational decision-making tends to be informed by policy rather than by local educational conditions or other shaping forces (Cremin, 1990, p. 72). Also, meganarratives typically overshadow small stories lived by K-12 students and their teachers in local settings. This chapter attempts to counterbalance "seeing [STEM] education small" – that is, at a distance through a systems lens (Greene, 1995) – as in longitudinal, multimillion dollar NSF grants and "seeing [STEM] education big" through up close, personal lenses of experiences that contributed to individual students' university and career trajectories. The act of "seeing [phenomena] big" in Greene's terms allows small stories of individuals to surface. At the same time, the ability to see both small and big helps researchers "to learn to move back and forth, to comprehend the domains of policy and long-term planning while also attending to particular [students], situation-specific undertakings, the unmeasurable, and the unique" (p. 11).

Description of Undergraduate and Graduate Students

Although many undergraduate and graduate students participating in the three scholarship grant programs stressed the influence their teachers had on their schooling, we decided to work with a convenience sample of three individuals with compelling stories to tell: Joyce Harding, Omid Kassem, and Lee Michaels (LM). The two males, Omid and LM, were students of color and the first in their families to attend university. As for Joyce, she is a rare white physics preservice candidate who is also the first in her family to enter higher education. Joyce, Omid, and LM each represent one of the three grant programs. Joyce Harding is a white female from Texas who was a participant in *teach*HOUSTON, the STEM teacher education grant; Omid Kassem is an Iranian American male who was part of the undergraduate cybertechnology program; and LM (Lee Mitchell) is an African American male who participated in the graduate cybersecurity program.

SMALL STORIES OF EXPERIENCE

Joyce Harding

Joyce Harding faced many challenges with her family during her K-12 years of schooling. She explained in her own words,

> I faced dark times growing up with my family… [I experienced] issues with [family members]. We were very poor and moved around a lot … and that definitely created a strain in my relationship with my parents. And it made it hard for me to want to do well in school…

> She added: "My family was a religious one and I can remember, even as a kid, my Dad saying things like 'Your Mother does not know anything because she comes from Eve'....and even as a young child, I thought there was something wrong with that argument..." In Joyce's opinion, what her parents taught her was "not right." Furthermore, they attempted "to suppress [her] propensity for critical thinking...." Consequently, when she began school, she encountered "...materials that would contradict things that [she] had learned at home."

Yet, the tough times Joyce experienced in her youth were often offset by "really great teachers who just gave me words of encouragement and advice that really stuck with me as I grew up." She continued: "I did have a few who did not really teach, but then I would always compare them to the many strong teachers I had who really taught and pushed us and asked deep probing questions... I soon recognized that was what good teachers did...." Still, Joyce admitted to having one or two teachers who were "awful to [her]" because "[her family's] electricity had been turned off [again] and [she] could not complete [her] assignments." At the same time, Joyce recognized that "the majority of [her] teachers were really understanding and would say: 'You know what? You are a good student' or 'What's going on in your life?' or 'I am not going to let this one grade affect you' or 'You can turn [the work] in when you get a chance'." They were "s-o-o-o understanding," Joyce observed, "and that meant the world to me." Hence, "over my childhood and then into my early teens, I learned how to handle knowledge from my teachers and that I could do my own thinking for myself."

Joyce eventually learned that logical fallacy was what her father was using when he traced her Mother's origins to Eve in a subtractive rather than an additive way. "I knew there was something awry with his argument. But I did not realize [as a young child] that it was ad hominin debate... that it was logical fallacy... That is what someone says when they do not have an argument."

The teachers who particularly helped Joyce seek out evidence to support viewpoints were her high school English teachers. According to her,

> One [male teacher particularly] caused me to question a lot of things. He made us read more advanced pieces such as *The Waste Land* by T. S. Eliot. Understanding that poem was very difficult. But I really, really liked the challenge... I like being intellectually stimulated and forced to ask hard questions... I like to think of things in different ways.

Hence, Joyce graduated from high school knowing that she was able "to ask better questions in class to get a better understanding of topics...." Further to this, she had learned from her primary and secondary years of schooling how to "access literature to help solve complex problems."

Later, when Joyce enrolled in undergraduate STEM classes and teacher education classes at university using her scholarship award, she encountered situations with a professor as well as with male students that echoed her experiences with her father in her home environment. She explained:

> One of my professors said that he expects males to do better in class because males have evolved to hunt and track the motions of animals and so they have better mental manipulation of objects and more abstract concepts... And females – because we have evolved to have babies – we are not so good at that...

From Joyce's perspective, "this was crazy [to say] in a gender-mixed classroom of over 100 students." Also, the professor did not select many females to answer questions in class. Joyce made the following sense of what was going on: "So, as a woman…in a lab, the professor ignores me because I am a girl…So, I started to do what the boys do. I started to call out answers."

Moreover, Joyce felt the professor's seeming gender bias helped to normalize sexist behaviors that may/may not have already existed among some male students in the laboratory. Here is how she explained one situation that transpired with her male peers and her:

> I was in a lab and the only female at a table and I was trying to assert that we can't solve two unknowns with only one equation…We needed to have a second equation. That is just basic algebra… I met serious opposition from the three males at the table who were sure I was wrong… I felt like [my solution] was dismissed…

At the same time, Joyce, in her words, "…looked around to see that were more males than females in the class – And then I remembered my test scores… They were higher than all of the guys around me, yet I was being ignored by the professor and the male students???' Immediately, Joyce exited the science lab and transferred from the male professor's lab to a similar STEM class taught by a diverse, female professor where no such issues existed. Joyce ended her small story by saying: "This is how the critical thinking skills that my teachers taught me in K-12 have continued to inform my education and actions and hopefully my future interactions."

Omid Kassem

Omid Kassem immigrated to the United States with his birth parents and his brother when he was a teenager. The family came from Iran to join his aunt who already resided in North America. Omid became an American citizen "as quickly as possible" after that. The move, he said, afforded "many opportunities" for his family. The fact that he could attend university tuition-free as the first member of his family was a "huge achievement." His father and mother – a grocery store clerk and a hair stylist in the Persian community – would never have been able to afford his tuition. Neither would they have been able to point him in the direction of higher education.

"It was not easy," said Omid, as he reflected backward on his high school education in the United States. This was mostly because he was a newcomer and "did not know how to speak English." In Iran, English is not taught as a school subject until high school. He explained: "High school was very hard for me…I had no knowledge of English grammar. I absolutely could not speak any English at all…." The following discussion about Omid's high school experience then ensued:

> Cheryl Craig: so you only attended high school in the U.S. for two years? Did you have English as a Second Language (ESL) classes in high school?
>
> Omid Kassem: Yes, but I was only in ESL for one semester because I needed to take two English courses [to fulfill my graduation requirements] …

Cheryl Craig: So that must have been tough?

Omid Kassem: Yes, it was very tough....

Cheryl Craig: You must have been a good student?

Omid Kassem: I graduated high school with a 3.9/4.0 Grade Point Average (GPA)...And, although I have been called talented in sports and some academic subjects, I think the term, *hard worker*, is a much better "label" for me.

Omid then proceeded to explain that he could only take two Advanced Placement (AP) classes because he did not have the Pre-AP prerequisites to enter other classes like Computer Science. The two AP classes he chose to enroll in were AP-Mathematics and AP-Science.

Omid Kassem, as readers probably have presupposed, was very strong in Mathematics and Science, scoring among the top 10%–15% in his home country before entering the US schooling system. However, he believes that he could never have achieved what he did without the support of his high school English as a Second Language (ESL) teacher who, fortuitously for him, spoke Arabic. It seems that she not only provided language support and ways to navigate his academic program in his high school but also helped him to be strong in a context entirely new to him and sometimes stacked against him as a newcomer to the school and/or the country. Not only did she stay after school to work one-on-one with Omid, she encouraged him to meet regularly with other students who were fluent in both English and Farsi to complete their homework assignments together. According to Omid,

> my ESL teacher was always supportive of me and encouraged me. She recognized my potential and kept in close touch with my other teachers through email...They told her how impressed they were with my progress in their classes, given how little language I knew.

However, his ESL teacher was not the only one who deserves special mention, according to Omid Kassem. Omid's success was also leveraged by his high school Science teacher who also played a leading role. That individual alerted him to the fact that the local, research intensive university had a Department of Computer Science that had scholarship grants available for talented students who were financially in need. Not only did that teacher encourage Omid to consider Computer Science, she mentioned the leader of the program's name and that professor would subsequently "take [him] under his wing." Since then, that professor has "provided me with an inside look at what computer science is" and "what it might be in my future." Omid further elaborated:

> ...My professor is so much more than a scholarship sponsor to me. He is my mentor and I go to him whenever I have questions about my courses, my projects and graduate school...He has taught me many things. He has cultivated my interest in research and pursuing a graduate degree. He's been such an influence that my project received the top prize in the summer of 2016.

In Omid Kassem's words, this is how teachers at his high school prepared him for university to work alongside the professor described above. This is how his teachers personally, socially, and academically helped him to enter the field of

Computer Science, a field that will enable him to obtain a high-salary, STEM position, which will lift his family's socioeconomic status while he concurrently serves as a role model for his brother.

Leon Mitchell (LM)

Leon Mitchell, who goes by LM, initially lived in a historical African American community where his parents had also been born and raised. In fact, his parents were part of a groundbreaking period in history: they were among the first African Americans bussed from their segregated black community to become racially integrated on white high school campuses. His mother graduated from the integrated high school setting; his father did not receive his high school diploma. His Mom immediately found secure employment cleaning offices in a high-rise building downtown, a job she continues to this day; his Dad worked as a laborer on various projects – sometimes employed, sometimes not. According to LM, his family is "a textbook case of the working poor." Still, the poverty his sister and he experienced "was similar to [their] classmates." Despite many low-income families living in their historical black neighborhood, "everyone in the small community lived in a circle of love."

In his early years of schooling, this feeling of comfort "while not necessarily having enough food" extended to the school. In his beginning years, "[LM's] kindergarten teacher and the school librarian were known for supplying small snacks to needy kids." LM continued: "Another go-to person was the choir director – which may have had something to do with the size of the school chorus!!!" He added that he always brought his younger sister to choir practice with him to make certain she made it to school on time, but also to guard against her being hungry. This was because their Mom left for work in the city center in the early hours of the morning and their father often did not come home.

In LM's late elementary years, however, his family arrangement changed. His father's alleged substance abuse problem had escalated to a point where his mother felt it "endangered the family." LM explained: "My Mom, my sister and I left in the middle of the night. From that point onward, my family – as I knew it – was never the same." His father visited LM and his sister intermittently but did not contribute regularly to the costs of their upbringing. In the meantime, his mother began to work overtime so she could provide for "[her children's] needs as best she could."

When his parents' marriage dissolved, LM and his sister not only lost daily contact with their Dad, they also lost the "security net" of their community – as LM referred to it. Furthermore, completing his final elementary years in school with mostly white classmates was "a culture shock" for him. Still, his sister and he continued to find teachers of all races who supplied them with food. One East Indian teacher, for instance, stood out in LM's telling. As he put it, "in whatever school I was in, there was someone who noticed that we were not always well-fed – and stepped up to the plate and made sure we were."

By middle school, LM's mother decided her family would return to a different part of the historical community where they previously had lived. This allowed her children to maintain contact with her former husband, extended family members, and friends. As he began middle school, LM said, in his words:

> I questioned what had happened to my family unit, the role of my Mother (who was not entirely innocent) and how my Dad had not seemed to shoulder his responsibilities… I was growing quickly and dealing with increased hunger…Also, it was "not cool" to have your little sister hanging around… I was becoming very angry and there were plenty of other angry kids with whom I could have gotten in trouble.

Trying not to assume a plotline like his father's while not fully succumbing to his Mother's matriarchal grip, LM continued to find emotional release in music and, by happenstance – a friend's referral – he discovered that participating in the track club helped him to develop laser-sharp focus. But it was not only running that assisted him in growing academically and physically. It was his African American high school track coach who became a "role model" to LM. What his coach did was regularly share breakfast one morning per week with his peers and him and engage them in conversations about sports, education, and life. One such discussion was particularly memorable:

> My track coach outright confronted me about my anger. He said that anger is an important stage in growing up for African American boys. He told me that I needed to channel my rage appropriately and to not let it destroy me… He said I must use it to make me stronger rather than weakening myself and risking becoming part of what is currently termed the school-to-prison pipeline… He held my shoulders, looked me directly in the eye and said that he thought I was smart and …capable of attending university in the future.

LM said the team then went out for a run. On that run, and on many more runs that followed, LM imagined how his future would unfold differently if he attended college and how that might affect not only his life trajectory, but the life trajectories of his sister, his mother – and even his father.

Two years later, LM finished middle school. At the graduation ceremony, he recalled his track coach shaking his hand, reminding him how he could be college-bound, and telling him that he had some contacts if LM ever decided to enroll in university. LM thanked him and pledged to keep in touch.

LM entered high school with thoughts of higher education floating in his head. In his classes, he continued to develop the mathematics, science, and fine arts focus offered by the magnet schools in his historically black community since his elementary years. As a high school freshman, he waved at his track coach a few times when he visited his old middle school to accompany his sister to appointments.

By the time LM was a junior, he knew he had the test scores, the participation in school and community activities, and the will to attend college. Still, he did not feel comfortable filling out college applications only with his high school counselor. He desired input from his track coach who had planted the seed of a university education in his middle school mind. Also, he vaguely remembered his coach saying he had some possible university pathways in mind for LM.

When LM returned to his former middle school to speak with his teacher-coach-mentor, this is what happened in LM's words:

> He [my track coach] did not seem fazed at all to see me… He quietly sat me down and said: "I knew you had it in you." He asked whether I was still interested in entering the sciences. I said yes – and that I was pretty certain I wanted to pursue an undergraduate degree in Computer Science. He inquired about the university campuses my high school counselor had suggested. He nodded quietly and then questioned: "Have you considered any historical black universities (HBUs)?" He said he knew a few – one in particular – that excelled in computer science education…

After this fruitful meeting with his track coach, LM returned to his high school counselor with three more colleges to add to his list. He then shared with the counselor that one of those three universities had become his first choice if a scholarship was to become available for him. After a very long wait, LM received positive feedback from three campuses, one being his first choice. He immediately sprinted from his home to his old middle school campus to share the good news with his track coach friend.

LM does not "know if the coach advocated for [him] behind the scenes…." What he does know is that he ended up with an esteemed African American professor as his undergraduate supervisor in Computer Science. That professor carefully guided him to the completion of his Bachelor of Science degree.[1] Furthermore, that individual knew that the Department of Computer Science at the comprehensive university near his historical African American community was one of the top 50 in the nation and that it had been awarded a prestigious NSF cybersecurity grant that would make graduate scholarships available to deserving students in need, providing they were American citizens. LM finished his small story by asserting that "an invisible web of teachers" supported him, eventually delivering him into the hands of professors who have done – and continue to do – the same.

SERIAL INTERPRETATION

Having featured three cases from the three awarded scholarship grants, we now present five overarching themes that cut across the representative cases. The five encompassing themes are: (1) same programs, different narrative histories; (2) in loco parentis, (3) learning in small moments; (4) counterstories; and (5) the importance of the liberal arts in STEM education.

Same Programs, Different Narrative Histories

When professors teach undergraduate classes sometimes numbering 100 students or more as Joyce Harding indicated, they tend to view students as a group, all being similar, all enrolled in similar classes and programs. However, our paper series shows the different ways that students' experiences have been shaped – or bent, as Schwab phrased it – by parents, teachers, and professors. Furthermore, even among these three possible effects, there are different ways that influence plays out on students. In this work, the longitudinal effect of undergraduate and graduate students' primary and secondary teachers is foregrounded. All three students, for one reason or another (high poverty, historically underserved minority, immigrant, ESL, etc.), would have been classified as likely failures in the state accountability system because it evaluates students' test results in relation to their peers' academic performances. At one point or another, each would have been "a bubble kid" – a student at high risk of dropping out. However, various teachers intervened in ways that greatly increased both the well-being and academic successes of Joyce, Omid, and LM. Think for a moment what could have happened without the contributions of their elementary and secondary teachers, for example, their home room teachers, the ESL teacher, the choir director, the track coach, the high school English teachers, their STEM teachers, and so forth. As Joyce succinctly put it, what her teachers did "meant the world" to her. Quite possibly Omid and LM would have said the same, if asked. Without their teachers' actions, their narratives – and, hence, their lives – would not have changed in the forward-moving ways they describe in this chapter.

In Loco Parentis

What teachers of all races did, of course, was to take the *in loco parentis* roles they played very seriously. *In loco parentis* means that teachers are charged with acting in students' best interests like Joyce's, Omid's, and LM's parents would have done were they able and present in the featured situations. In this chapter, we see teachers acting as proxies for parents by providing students with food, listening to their stories, engaging them in hard conversations, offering advice when needed, and providing words of comfort, support, and direction. The teachers also openly advocated for Joyce, Omid, and LM. The teachers' actions clearly aided the students' best-loved selves (Craig, 2013; Schwab, 1954/1978). We would argue that these teachers were not personally enacting "teacher as savior" roles (Brown, 2013; Trop, 1996, as cited in Ladson-Billings, 1998) through "rescuing" students from themselves, their families, and their communities (Baker, 1996, as cited in Ladson-Billings, 1998). Rather, what these teachers did – beyond being responsible professionals – was what "the best and wisest parent[s]" would do on behalf of their own children (Dewey, 1899/1980, p. 5). At the same time, the American educational system – through denying sufficient resources to the schools and special assistance to at-risk learners – leaves teachers in impossible positions, according to António Nóvoa (2013) who explained that "problems…will not be solved only within [schools]. Prophecies of salvation through [schools] enclose teachers in unreasonable ambitions, blaming them for the failure of school reform" (p. 35).

Counterstories

The teachers – frequently acting as parental proxies for Joyce, Omid, and LM – planted seeds for different "stories for them to live by," but the students themselves had to accept the plotlines of these counternarratives as part of their identities. For Joyce, her counterstory had to do with learning to think and reason for herself despite the "truths" she had learned in her family. For Omid, a first-generation-to-college youth, it meant learning ESL, Mathematics, and Science at a highly accelerated rate. This enabled him to take full advantage of the opportunities that the American educational system offered. For LM, it involved seeing himself as university-bound even when that story did not exist in his family or in his historically black community. In all three students' cases, their teachers introduced them to powerful counternarratives to live by, which the students embraced through breathing self-sustaining life into them and making them their own. As Lindemann Nelson (1995) explained: "counterstories...are told in dialogue with others…they are …told together with other tellers, fragment by fragment, each person contributing to plot and character and…thought" (p. 38). This counterstorying process is especially evident in the point–counterpoint conversations that took place between LM and his track coach.

Learning in Small Moments

Small stories unfurling over time between many teachers and the three students helped the youths to excel in the STEM disciplines in their undergraduate and graduate education and fueled their hopes for future STEM-related careers. These stories, which sustained Joyce, Omid, and LM and prepared them for higher education, were not grand stories or meganarratives (Olson & Craig, 2009) of STEM education. Rather, they were small stories – petite plotlines – that emerged in rather mundane situations in everyday elementary and secondary school life. In fact, for the students, it was all about "learning in small moments" (Meier, 1997). These "bite-sized" narratives, as Meier called small stories, laid the foundation for the stunning progress the students made. For Joyce who was enrolled in *teach*HOUSTON, it was a constellation of K-12 teachers who helped her become an independent thinker, always weighing evidence before arriving at decisions. For Omid, it was enrolling in Science as a high school option and having a dedicated teacher who knew that the local university had a scholarship program in Computer Science for which he would qualify. Finally, for LM, it was eating breakfast weekly with his middle school track coach and other teammates and discussing sports, education, and life, which subsequently opened the doors to his majoring in a STEM discipline.

Importance of the Liberal Arts in STEM Education

The teachers who supported Joyce, Omid, and LM on their journeys to undergraduate and graduate education in the sciences were not only STEM educators but also liberal arts (including performing arts) teachers. Martha Nussbaum (2010) asserts that the lessons that the humanities teach about emotions, human differences, and community are critical to the maintenance of democracy at a time when it is fiercely under attack by neoliberals calling for the marketization and privatization of public education (i.e., Cochran-Smith, 2003). Nussbaum maintains that those who teach the humanities teach students about human emotions and human frailties in ways that those who teach the STEM disciplines are not able. In the featured narratives, we can specifically connect Joyce's

English teacher and Omid's ESL and Science teachers with LM's track coach and choir director since "physical education can be identified as one of the forms of liberal education, comparable with music and the non-verbal arts as a source of one of the kinds of meaning that enriches man's [*sic*] comprehension of reality as he knows it" (Methany & Ellfeldt, 1961, pp. 282–290). In fact, LM's small story uncannily resembles that of philosopher Michael Oakeshott (1989) who credited his gymnastic teacher with introducing him to "patience, accuracy, economy, elegance, and style" (p. 62).

CLOSING STATEMENTS

What this chapter on the influence of teachers on primary and secondary students who enroll in university-based STEM programs in the hope of entering discipline-related professions shows is that the public sadly takes too narrow of a view of teachers, their roles in a democratic society, their input into students' experiences, engagement, and understandings, and their impact on students' attitudes, achievement, and attainment. In the local school system, the value of a teacher is currently reduced to one academic year of influence via value-added modeling approaches and is considered through the lens of one content area in the middle school and high school years. This exploratory research study abundantly shows the influence of teachers acting as agents of education in its entirety, not simply as agents in one disciplinary field.

It is therefore essential that teacher education programs prepare prospective teachers for their comprehensive roles, which necessarily includes *in loco parentis* responsibilities. Furthermore, the concerns of the unknown teachers featured in this chapter were not merely for the academic development of Joyce, Omid, and LM, but also for their physical, emotional, and social well-being – in short, their teachers actively assisted these students in developing a strong sense of their best-loved selves. Their teachers' understanding of their roles as pedagogues involved the development of Joyce, Omid, and LM as whole persons, a perspective that necessarily needs to be stressed in teacher education programs despite it being an unmeasurable quality. Additionally, the public tends to look at what is calculable, observable, and marketable about students from an outside perspective but fails to see "the skein of myriad threads that know no boundaries" and how these threads intersect in highly productive, life-giving ways in students' inner beings. This fine-grained research demonstrates how qualitative studies can be productively accumulated across grant programs. It vividly shows how the concerted efforts of "an army of unsung teachers" has had a lasting impact on the disciplines the particular students choose to study, the careers to which they will devote themselves, and the lives they wish to live. Lastly, Joyce's, Omid's, and LM's narratives of experience offer insights into what might constitute an ideal STEM teacher education curriculum, a curriculum that would be a far cry from the neoliberal version currently being touted. Finally, the university students' stories – alone and together – form models that are instructive to their siblings and other STEM students who will follow in their footsteps as well as to teachers

(Heilbrun, 2008) who will feel heartened and affirmed to hear of their lasting human impact, especially when societal forces demonize public education and minimize, even nullify, teachers' contributions.

AUTHORS' NOTE

This research was funded by the following NSF Grant Programs:

- Noyce Grant: *Recruitment, Preparation and Retention of STEM Students as High School Teachers* [1240083]
- S-STEM Scholar Program: *Preparing Computer Science Students for Global Challenges of the twenty-first Century* [1356705]
- CyberCorps (R): *Scholarship For Service (SFS): Increasing Talented Trusted Professionals* [1433817]

NOTES

1. The influence of professors is the next chapter in this book.

REFERENCES

Bateson, M. C. (1994). *Peripheral visions: Learning along the way*. New York, NY: Harper Collins.

Brown, A. (2013). Waiting for superwoman: White female teachers and the construction of the "neoliberal savior" in a New York City public school. *Journal for Critical Education Policy Studies, 11*(2), 123–164.

Bruner, J. (2002). *Making stories: Law, literature and life*. New York, NY: Farrar, Strauss and Giroux.

Bruner, J. (2010). Narrative, culture, and mind. In D. Schiffrin, A. De Fina, & A. Nylund (Eds.), *Telling stories: Language, narrative, and social life* (pp. 45–49). Washington, DC: Georgetown University Press.

Caine, V., Murphy, M. S., Estefan, A., Clandinin, D. J., Steeves, P., & Huber, J. (2016). Exploring the purposes of fictionalization in narrative inquiry. *Qualitative Inquiry, 23*(3), 215–221. doi: 10.1177/1077800416643997.

Chang, P., & Rosiek, J. (2003). Anti-colonial antinomies: A case of cultural conflict in the high school biology curriculum. *Curriculum Inquiry, 33*(3), 251–290.

Ciuffetelli Parker, D., & Craig, C. (2017). An international inquiry: Stories of poverty–poverty stories. *Urban Education, 52*(1), 120–151.

Clandinin, D. J. (2013). *Engaging in narrative inquiry* Walnut Creek, CA: Left Coast Press.

Clandinin, D. J., & Connelly, F. M. (1995). *Teachers' professional knowledge landscapes*. New York, NY: Teachers College Press.

Clandinin, D. J., & Connelly, F. M. (2000). *Narrative inquiry: Experience and story in qualitative research*. San Francisco, CA: Jossey-Bass.

Clandinin, D. J., Murphy, M. S., Huber, J., & Orr, A. M. (2009). Negotiating narrative inquiries: Living in a tension-filled midst. *The Journal of Educational Research, 103*(2), 81–90.

Clandinin, D. J., & Rosiek, J. (2006). Mapping a landscape of narrative inquiry: Borderland spaces and tensions. In D. J. Clandinin (Ed.), *Handbook of narrative inquiry: Mapping a methodology* (pp. 35–75). Thousand Oaks, CA: Sage Publishing.

Cochran-Smith, M. (2003). Sometimes it's not about the money: Teaching and heart. *Journal of Teacher Education, 54*(5), 371–375.

Cohen, D., & Garet, M. (1975). Reforming educational policy with applied social research. *Harvard Educational Review, 45*(1), 17–43.

Coleman, J. S., Campbell, E. Q., Hobson, C. J., McPartland, J., Mood, A. M., Weinfeld, F. D., & York, R. L. (1966). *Equality of educational opportunity (No. FS 5.238:38001)*. Washington, DC: National Center for Educational Statistics, Office of Education, U.S. Department of Health, Education, and Welfare.

Connelly, F. M., & Clandinin, D. J. (1990). Stories of experience and narrative inquiry. *Educational Researcher*, *19*(5), 2–14.

Connelly, M., & Clandinin, D. J. (1986). On narrative method, personal philosophy, and narrative unities in the story of teaching. *Journal of Research for Science Teaching*, *23*(4), 293–310. doi: 10.1002/tea.3660230404

Connelly, M., & Clandinin, D. J. (1999). *Shaping a professional identity*. New York, NY: Teachers College Press.

Connelly, M., & Clandinin, D. J. (2005). Narrative inquiry. In J. Green, G. Camilli, & P. Elmore (Eds.), *Handbook of Complementary Methods in Educational Research* (pp. 477–489). Washington, DC: American Educational Research Association.

Craig, C. (2007). Story constellations: A narrative approach to situating teachers' knowledge of school reform in context. *Teaching and Teacher Education*, *23*(2), 173–188.

Craig, C. (2011). Narrative inquiry in teaching and teacher education. In J. Kitchen, D. Ciuffetelli Parker, & D. Pushor (Eds), *Narrative inquiries into curriculum making in teacher education* (pp. 19–42). Bingley: Emerald Group Publishing Limited.

Craig, C. (2013). Teacher education and the best-loved self. *Asia Pacific Journal of Education*, *33*(3), 261–272.

Craig, C. (2018). Metaphors of knowing, doing and being: Capturing experience in teaching and teacher education. *Teaching and Teacher Education*, *69*, 300–311.

Craig, C., Evans, P., Stokes, D., Bott, S., & Abrol, B. (2017). Attracting, preparing and retaining teachers in high need areas: A science as inquiry model of teacher education. In M. Peters, B. Cowie, & I. Menter (Eds.), *A companion for teacher education research*. Dordrecht: Springer Publications.

Craig, C., Verma, R., Stokes, D., Evans, P., & Abrol, B. (2018). The influence of parents on undergraduate and graduate students' entering the STEM disciplines and STEM careers. *International Journal of Science Education*, *40*(6), 621–643.

Craig, C., You, J., Zou, Y., Verma, R., Stokes, D., Evans, P., & Curtis, G. (2018). The embodied nature of narrative knowledge: A cross-study analysis of embodied knowing in teaching, learning and life. *Teaching and Teacher Education*, *71*, 329–340.

Cremin, L. (1990). *Popular education and its discontents* New York, NY: Harper & Row Publishers.

Darling-Hammond, L. (2000). Teacher quality and student achievement: A review of state policy evidence. *Education Policy Analysis Archives*, *8*(1), 1–44.

Dewey, J. (1899/1980). *The school and society*. Carbondale, IL: Southern Illinois University Press.

Dewey, J. (1916). *Democracy and education: An introduction to the philosophy of education*. New York NY: Macmillan.

Dewey, J. (1938). *Experience & education*. New York, NY: Kappa Delta Pi.

Dietz, C., & Davis, E. (2009). Preservice elementary teachers' reflection on narrative images of inquiry. *Journal of Science Teacher Education*, *20*(3), 219–243.

Georgakopoulou, A. (2004). Narrative analysis workshop: How to work with narrative data. Paper presented at narrative matters 2004: An interdisciplinary conference on narrative perspectives, approaches, and issues across the humanities and social sciences, Fredericton, NB, May 20–23.

Goldhaber, D. (2016). In schools, teacher quality matters most. *Education Next*, *16*(2), 56–62. Retrieved from http://educationnext.org/in-schools-teacher-quality-matters-most-coleman/

Greene, M. (1995). *Releasing the imagination: Essays on education, the arts, and social change*. San Francisco, CA: Jossey-Bass.

Hattie, J. (2003, October). Teachers make a difference: What is the research evidence? Paper presented at the Building Teacher Quality: What does the research tell us ACER research conference, Melbourne, VIC. Retrieved from http://research.acer.edu.au/research_conference_2003/4/

Hattie, J. (2008). *Visible learning: A synthesis of over 800 meta-analyses relating to achievement*. Abingdon-on-Thames: Routledge.

Heilbrun, C. (2008). *Writing a woman's life*. New York, NY: W.W. Norton & Company.

Hollingsworth, S., Dybdayl, M., & Minarik, L. T. (1993). By chart and chance and passion: The importance of relational knowing in learning to teach. *Curriculum Theory*, *23*(1), 5–35.

Jencks, C., Smith, M., Acland, H., Bane, M., Cohen, D., Gintis, H., ... Michelson, S. (1972). *Inequality: A reassessment of the effect of family and schooling in America*. New York, NY: Basic Books.

Jordan, H., Mendro, R., & Weersinghe, D. (1997). Teacher effects on longitudinal student achievement: Preliminary report on research on teacher effectiveness. Paper presented at the National Evaluation Institute, Indianapolis, IN. Kalamazoo, MI: CREATE, Western Michigan University.

King, T. (2003). *The truth about stories*. Toronto, CA: Anansi Press.

Kurth, L. A., Kidd, R., Gardner, R., & Smith, E. L. (2002). Student use of narrative and paradigmatic forms of talk in elementary science conversations. *Journal of Research for Science Teaching*, *39*(9), 793–818. doi:10.1002/tea.10046

Ladson-Billings, G. (1998). Teaching in dangerous times: Culturally relevant approaches to teacher assessment. *The Journal of Negro Education*, *67*, 255–267.

Lindemann Nelson, H. (1995). Resistance and insubordination. *Hypatia*, *10*(2), 23–40.

Lyons, N. (2010). Narrative inquiry: What possible influences on policy or practice? In D. J. Clandinin (Ed.), *Handbook of narrative inquiry: Mapping a methodology* (pp. 600–631). Thousand Oaks, CA: Sage Publishing.

Meier, D. R. (1997). *Learning in small moments: Life in an urban classroom*. New York, NY: Teachers College Press.

Methany, E., & Ellfeldt, L. (1961). An inquiry into the nature of movement as a significant form of human experience. In E. Davis (Ed.), *The philosophic process in physical education* (pp. 282–290). Philadelphia, PA: Lea & Febiger.

National Commission on Teaching and America's Future. (1996, September). *What matters most: Teaching for America's future*. New York, NY: Author.

Nóvoa, A. (2013). Teachers: How long until the future? In M. Assunção Flores, A. Carvalho, F. Ilídio Ferreira, & M. Vilaça (Eds.), *Back to the future: Legacies, continuities and changes in educational policy, practice and research* (pp. 29–37). Rotterdam: Sense Publishing.

Nussbaum, M. (2010). *Not for profit: Why democracy needs the humanities* (Vol. 2). Princeton, NJ: Princeton University Press.

Oakeshott, M. (1989). *The voice of liberal learning: Michael Oakeshott on education*. New Haven, CT: Yale University Press.

OECD. (2005). *Teachers matter: Attracting, developing and retaining effective teachers*. Retrieved from http://www.oecd.org/education/school/34990905.pdf

Olson, M. (2000). Where the story leads: A response to Cheryl Craig. *Canadian Journal of Education*, *25*(4), 349–352. doi:10.2307/1585856

Olson, M., & Craig, C. (2009). "Small" stories and meganarratives: Accountability in balance. *Teachers College Record*, *111*(2), 547–572.

Palmer, P. (1998). *The courage to teach: Exploring the inner landscape of a teacher's life*. San Francisco, CA: Jossey-Bass.

Pushor, D. (2013). Teachers' narrative understandings of parents: Living and reliving "Possible Lives" as professionals. *Journal of Family Diversity in Education*, *1*(1), 40–57.

Roorda, D. L., Koomen, H. M., Spilt, J. L., & Oort, F. J. (2011). The influence of affective teacher–student relationships on students' school engagement and achievement: A meta-analytic approach. *Review of Educational Research*, *81*(4), 493–529.

Ross, V. (2003). Walking around the curriculum tree: An analysis of a third/fourth grade mathem(atic)s lesson. *Journal of Curriculum Studies*, *35*(5), 567–584. doi:10.1080/0022027032000083560

Ross, V. (2004). A story of reform: Math, science, technology investigation (MSI) in room 34 at Bay Street community school. *Journal of Curriculum Studies*, *36*(5), 587–594. doi:10.1080/0022027032000167000a

Sack, J. (2008). Common intersections within a high school Mathematics Leadership Institute. *Journal of Teacher Education*, *59*(2), 189–199. doi:10.1177/0022487107314003

Sack, J., Quander, J., Redl, T., & Leveille, N. (2016). The community of practice among mathematics and mathematics education faculty members at an Urban Minority Serving Institution in the U.S. Innovative *Higher Education*, *41*(2), 167–182. doi:10.1007/s10755-015-9340-9

Sanders, W., & Rivers, J. (1996). *Cumulative and residual effects of teachers on future student academic achievement*. Knoxville, TN: University of Tennessee Value-Added Research and Assessment Center.

Schwab, J. (1954/1978). Eros and education: A discussion of one aspect of discussion. In I. Westbury & N. Wilkof (Eds.), *Science, curriculum and liberal education: Selected essays*. Chicago, IN: University of Chicago Press.

Schwab, J. (1956/1978). Inquiry and the reading process. In I. Westbury & N. Wilkof (Eds.), *Science, curriculum and liberal education: Selected essays*. Chicago, IL: University of Chicago Press.

Schwab, J. (1960). Enquiry, the science teacher, and the educator. *The Science Review*, *68*(2), 176–195.

Schwab, J. (1961/1978). Education and the structure of the disciplines. In I. Westbury & N. Wilkoff (Eds.), *Science, curriculum and liberal education: Selected essays*. Chicago, IL: University of Chicago Press.

Schwab, J. (1980). Ends and beginnings. In H. Gideonse, R. Koff, & J. Schwab (Eds.), *Values, inquiry, and education*. Los Angeles, CA: Center for the Study of Evaluation, University of California.

Schwab, J. (1983). The practical 4: Something for curriculum professors to do. *Curriculum Inquiry*, *13*(3), 239–265.

Schwab, J. J. (1969). *College curriculum and student protest*. Chicago, IL: University of Chicago.

Schwab, J. J. (1973). The practical 3: Translation into curriculum. *School Review*, *83*, 501–522.

Schwartz, K. (2015). *Could storytelling be the secret sauce of STEM education?*. MindShift. Retrieved from https/www.kqued.org/mindshift/39949/could-storytelling-be-the-secret-sauce-to-stem-education

Stone, E. (1988). *Black sheep and kissing cousins: How our family stories shape us*. New York: Times Books.

UNESCO. (2011). *UNESCO ICT competency framework for teachers*. Retrieved from http://nesdoc.unesco.org/images/0021/002134/213475e.pdf

UNICEF. (2009, March). *Child friendly schools manual*. Retrieved from http://www.unicef.org/publications/files/Child_Friendly_Schools_Manual

Varelas, M., Kane, J. M., & Wyle, C. D. (2012). Young black children and science: Chronotopes of narratives around their science journals. *Journal of Research for Science Teaching*, *49*(5), 568–596. doi:10.1002/tea.21013

Wagler, R. (2010). Using science teaching case narratives to evaluate the level of acceptance of scientific inquiry teaching in preservice elementary teachers. *Journal of Science Teacher Education*, *21*(2), 215–226. doi:10.1007/s10972-009-9160-9

Wright, S., Horn, S., & Sanders, W. (1997). Teacher and classroom context effects on student achievement: Implications for teacher evaluation. *Journal of Personnel Evaluation in Education*, *11*, 57–67.

Xu, S., & Connelly, F. M. (2010). Narrative inquiry for school-based research. *Narrative Inquiry*, *20*, 349–370.

THE INFLUENCE OF PROFESSORS ON STUDENTS ENROLLED IN THE STEM PROGRAMS WITH THE INTENT OF EMBARKING ON STEM CAREERS

Jing Li, Paige K. Evans, Cheryl J. Craig,
Donna W. Stokes, Rakesh Verma and Gang Zhu

ABSTRACT

Scant attention has been paid to the influence of professors on science, technology, engineering, and mathematics (STEM) students' learning and lives at the tertiary level. To fill this void, this chapter examines the influence of professors on students' entering and remaining in the STEM disciplines and pursuing STEM careers within the context of six funded STEM grants in the southern United States. We examine professor–student interactions using the students' storied experiences as the fodder for our narrative inquiry. We present narrative exemplars from which the following themes emerged: (1) agency as a student and agency as a human being, (2) development of students' multilayered identities, and (3) professors' engagement of themselves in their interactions with students. A discussion of learner-centeredness and professors' professional development in higher education concludes this study of professors' influence on students' learning and intended careers.

Keywords: Higher education; professor influence; student learning and lives; learner-centeredness; STEM disciplines; STEM careers; *teach*HOUSTON

Concerns have been raised across many nations about workforce needs in the science, technology, engineering, and mathematics (STEM) fields (Marginson,

Preparing Teachers to Teach the STEM Disciplines in America's Urban Schools
Advances in Research on Teaching, Volume 35, 159–177

ISSN: 1479-3687/doi:10.1108/S1479-368720210000035010

Tytler, Freeman, & Roberts, 2013). In the United States where our study is located, the STEM workforce shortage is likely to reach two million over the next decade if some kind of intervention does not happen (Giffi et al., 2015). Post-secondary education ultimately prepares students for their STEM careers. However, less than half of the three million US college students intending to major in STEM fields annually persist in their STEM programs of study until graduation (PCAST, 2012). Sustaining STEM students in their majors thus becomes an urgent concern of higher education. Internationally, "the single most significant means of improving education…is through teaching" (Ministry of Education of China, 1993; Pollard 2010, p. 1). Time and time again, it has been confirmed that nothing is more important to students than the people who teach them (Schleicher, 2018). The close connections between student success and those who teach them suggest professors have a crucial role to play in higher education just as teachers play an essential role in schools.

Against the interwoven national and international backdrops of teacher–student relatedness, postsecondary education, and STEM workforce needs, our study queries the influence of professors on undergraduate and graduate students enrolled in STEM programs who plan to enter into STEM careers in the United States. In our study, "professor" is a term generally referring to anyone (adjunct/untenured/tenured of any rank) who teaches and mentors students in higher education, which distinguishes itself from the word "teacher" that is conventionally used at K-12 contexts. Based on the student interviews and focus group sessions in our ongoing research of six STEM scholarship grant programs funded by the National Science Foundation (NSF) in the United States, our research team narratively disentangles – from the undergraduate and graduate students' perspectives – how their professors helped the individual students to grow personally and academically so they can thrive in their intended careers as STEM professionals. Our research also provides insights into the learner-centeredness and professors' professional development in universities. All those ultimately contribute knowledge to the retention of STEM students and teaching/mentoring in higher education as a whole, which are critically important both now and in the future, nationally and internationally.

LITERATURE REVIEW

STEM Student Attrition and Retention

As foreshadowed, low student enrollment and high attrition rates in STEM education present formidable challenges in higher education (Sithole et al., 2017). Exit rates are especially high for the "underrepresented majority," which includes women, and racial/ethnic minorities enrolled in STEM majors (Broecke & Hamed, 2008). Factors affecting the underrepresentation of women and students of color include less positive racial climate, negative societal stereotypes, marginalization, and microaggressions (Chang, Eagan, Lin, & Hurtado, 2011; Dortch & Patel, 2017; D. R. Johnson, 2007). Those may give rise to loneliness, self-doubt, and ultimately student attrition from the STEM majors.

Current studies have explored STEM students' retention from varied lenses. For instance, Freeman, Alston, and Winborne (2008) found that peer learning community engages students in active, collaborative learning and identity making as scientists. Also, supportive educational environments during college are positively linked to the retention of students of color in STEM majors (Bonous-Hammarth, 2000). Studies in this field mainly seek answers to student attrition from the perspectives of students' knowledge construction, intellectual growth, and professional identity. Our study, based on students' organic narratives of their interactions with professors, reports how the professors' beings and practices could sustain STEM students and lead them to leave as well.

The Influence of Professors

In higher education, student–faculty interactions in formal or informal occasions are critical experiences associated with student development (Cotton & Wilson, 2006). The behaviors of female and male faculty members may have different indications for students and in turn impact their learning (Smeby, 2000). The professors' narrow focus on science only as a decontextualized subject matter would discourage students and detract from their overall college experiences (A. Johnson, 2007). Smith and Allen (2006) also discovered that student-centered mentoring helps students to integrate life, career, and educational goals in a more satisfying way. Furthermore, faculty can impact students' aspirations to pursue further degrees (Trolian & Parker, 2017). Although there is less study on student–faculty interactions in the STEM fields, it is commonly recognized that relationships with professors influence STEM students' learning performance and retention (Xu, 2016).

Most of the aforementioned research document professors' influences on students through macrocategories abstractly defined. Our study bridges the gap through narratively presenting the nuanced ways that professors influence students' learning, life, and decision-making in their journeys to becoming STEM professionals. Furthermore, we also illuminate our multilayered understandings of how professors can potentially address the disjunction between the sufficient opportunities for students to enter STEM careers and students' lack of confidence and enthusiasm about embarking in STEM fields of study.

THEORETICAL UNDERPINNINGS

In this section, we introduce three theoretical pillars that underpin our presentation and interpretation of students' experiences with their professors: experience and story; identity; and agency, which will segue readers into our full-blown discussion of the topic at hand.

Experience and Story

Dewey (1938) emphasized the organic connections between experience and education, which enabled people to understand educational life in a new, experiential sense. Experience is "a changing stream...characterized by continuous interaction of human thought with [the] personal, social and material environment" (Clandinin & Rosiek, 2007, p. 39). The meanings within experience need to be explored because experience is not "self-explanatory" (Dewey, 1938, p. 25). As Paul Ricoeur (2003a, p. 27) confirmed: "A life is no more than a biological phenomenon as long as it has not been interpreted." Further to this, experience is best understood through storytelling and narratives (Craig, Evans, Verma, Stokes, & Li, 2018). Therefore, presenting student–professor interactions as storied experiences enables us to interweave characters, emotions, and circumstances in a way that further helps us to retrieve multiperspectival meanings regarding how professors influenced the students' experiences in higher education and potentially shaped their personal and professional beings in the future. This narrative view of experience also lays a solid theoretical foundation for the narrative inquiry method we use in our study, which will be discussed in the methodology section.

Identity

It is important that we can "identify ourselves" with the present person we are and the future person we want to be before we "integrate our actions" and make plans for the future (Mackenzie, 2008, p. 9). Approaching students' identity from an impersonal lens technically equates a man's brain to a man as a person (Ricoeur, 2003b). To the members of our research team, identity is "a relational phenomenon" (Beijaard, Meijer, & Verloop, 2004, p. 108) that plays out socially and individually in the continuum of experience. This strongly echoes Ricoeur's (1988) elaborations on identity that people have to continuously negotiate the relationship between the third-personal and first-personal dimensions of identity on the time continuum of past–present–future. Hence, identity is always in the making and never made (Greene, 1995) in the changing milieus.

With this experiential, relational, and dynamic view of identity, we probe the complexities of student identity formation and shifting which become apparent in their "contingent, subjective, and momentary emergence" (Malone & Barabino, 2009, p. 489) in student–professor interactions. In a nutshell, through "an ongoing process of interpretation and reinterpretation of experiences" (Beijaard et al., 2004, p. 122), we capture how students, under the influence of their professors, decided "how to be," "how to act," "how to understand" (Beauchamp & Thomas, 2009, p. 178), and how to find their own fulfillment (Taylor, 1991/2003, p. 29), in their learning, in their unfolding lives (King, 2003), and in their future careers.

Agency

Embedded within the experiential view of identity is a narrative view of one's self that plays a central role in understanding human agency (Kennett & Matthews, 2008). Humans are a "self-moving living thing" (Schwab, 1964, p. 8) who "has a notion of the future and the past, can hold values, make choices; in short, can adopt life-plans" (Taylor, 1985c, p. 97), which is conceptualized by Taylor as part of self-evaluation.

According to Taylor (1985a, pp. 16–18), a human with agency can do "a qualitative evaluation" of one's desires – distinguish desires "as to worth," which is named as "strong evaluation". Different from the choices one simply makes based on preferences and consequences, strong evaluators concern about "qualitatively different modes of life: fragmented or integrated, alienated or free, saintly, or merely human…" (Taylor, 1985a, p. 16). Therefore, a human with agency involves a "reflexive being" (Taylor, 1985b, p. 60) who has self-initiated views of the kind of life one wants to live, which together with experiences constitute what people are.

Our study examines how the students, with the help of their professors, enhanced their autonomy, nourished their inwardness (Taylor, 1991/2003), and opened to new matters significant to them (Taylor, 1985c). In sum, the theoretical pillars interweave with each other and underpin our narrative inquiry into how the STEM students, under the influences of their professors, draw a map (Taylor, 1985b) of their own life and career trajectories.

RESEARCH METHOD

Narrative Inquiry

Narrative inquiry is a qualitative method that fluidly studies experiences as story "over time, in a place or series of places, and in social interactions with milieus" (Clandinin & Connelly, 2000, p. 20). We thus consider students as "individuals as living storied lives on storied landscapes" (p. 24). Reflecting Dewey's (1938) philosophy of experience, our narrative inquiry is unfolded in a "three-dimensional narrative inquiry space" (Clandinin & Connelly, 2000, p. 54) as shown in Fig. 1. We unfurl students' experiences on a past–present–future continuum (continuity) and explore students' communications with professors/others (interactions) inside and outside classrooms (situation).

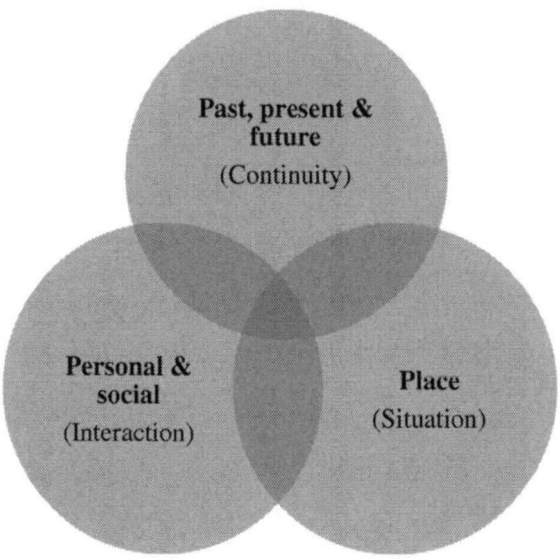

Fig. 1. Three-dimensional Narrative Inquiry Space.

Three analytical tools of narrative inquiry – broadening, burrowing, and storying and restorying (Connelly & Clandinin, 1990) – enable us to travel in the three-dimensional space and seam together the storied data we gathered from interviewing research participants in the six grant programs: (1) broadening positions our study in the national and international urgent workforce needs in the STEM fields and the usually taken-for-granted nature of professor–student interactions in higher education; (2) burrowing reveals what bubbled to the surface in our distinct and combined data pools of students' storied experiences with their professors; (3) storying and restorying allow us to track how students, impacted by their professors, claimed their agency and altered the plotlines of the stories they live and tell. Together, the three interpretive devices guide us to think seriously about how the meanings embedded in the individual narratives relate to the larger world and capitalize on possibilities in professor–student interactions in higher education, rather than simply looking for "full-fledged stories" (Bamberg & Georgakopoulou, 2008, p. 377).

Validation

The quality of our narrative research is ensured by a process of validation which involves the principles of reflexivity, historical continuity, and dialectics (Heikkinen et al., 2007, 2012). With regard to the reflexivity, we have presented the conceptual framework that shaped our interpretation of students' experiences so that both readers and we as researchers are aware of the way of knowing in this study. Where the principle of historical continuity is concerned, we will carefully illuminate how different students' experiences with their professors evolved on the past–present–future continuum. As for the principle of dialectics, we member-checked (Lincoln & Guba, 1985) with the student participants and sought multiple types of informal feedback from them to heighten the interpretive level of this chapter and the resonances students found with it. Further to this, part of the validation relies on readers. As Ricoeur (2003a, p. 26) emphasized, "the world of the reader and the world of the text" need to "fuse together" so that the narrative establishes its significance in readers' reinterpretations and imagined use of the discussed practices in their own milieus.

Research Participants

The six NSF grants are awarded to the University of Houston (UH), the second most diverse, research-intensive campus in the United States. Two of them are S-STEM grant programs, which, respectively, prepare computer science students and physics students for the twenty-first century global challenges through a multidisciplinary approach. Another three grants are Robert Noyce Scholar Grants awarded to *teach*HOUSTON program (a designer teacher education program), with the purpose of supporting mostly undergraduate mathematics and science majors to teach at high-needs secondary schools. The last but not the least is Scholarship for Service grant. It provides scholarships for students' education, training, and opportunities to become cybersecurity professionals.

All participants in the six grant programs engaged in multiple interviews and focus groups sessions as well as email communications while supported by the NSF-funded scholarships. For this chapter, we chose a sample of convenience (Marshall, 1996) of undergraduate/graduate students and explored what they had to say about the influence of their professors on them. In our data pools, we found evidence clustering around many instructors. Our research team decided to create five narrative exemplars (Lyons & LaBoskey, 2002; Mishler, 1990) capturing the most common themes that represented all the grant programs with some students being participants in more than one NSF initiative. Some narratives we feature involve multiple students' storied experiences with one professor. Other narratives illuminate one student's stories with one or multiple professors. Thus, the five narrative exemplars together draw students and professors into a mix, present varied socially and contextually embedded situations, and provide multiperspectival insights into how professors impacted students' identities, lives, and careers.

Additionally, we did not assign pseudonyms to all of our student participants. Only students named in previous publications have pseudonyms. Also, if the instructors in our data pools were individual professors, we assigned them fictitious names. We wanted to streamline how our research participants were presented to our readers so they could place full attention on undergraduate/graduate students' storied experiences of their professors.

Narrative Exemplars

The small stories of experience we now narratively offer as exemplars convey five ways that STEM professors influenced the undergraduate/graduate students participating in the six grant programs. The five narratives congeal around the following themes: (1) the professor cares, (2) the professors build a STEM community, (3) the professor represents females in the STEM disciplines, (4) the professor acts as a navigator for students, and (5) the professors take an inquiry stance to teaching and learning.

The Professor Cares

When asked in passing whether there were professors in their STEM programs who were influencing them, some students pinpointed Dr. Patel in our interview and focus group discussions. One student said the following about Dr. Patel's computer science courses:

> My favorite class has to be… data structures. And my favorite professor is also teaching it. I have had her for two semesters. She really cares about her students. She does not, like, just show up to work just to read… PowerPoint slides. She actually *wants* you to ask questions. She *wants* you to participate. She *wants* you to present your projects to her. (italics indicates words receiving emphasis)

The students were also impressed by the extra credit assignments that Dr. Patel made available to them. They said they did not need to complete these assignments, but many chose to do so because they "buy into [Dr. Patel] as a person and as a professor." Moreover, for those students, the assignments were not just one-off tasks Professor Patel used to give them extra points, but activities that "challenged" them to think and provided a "safe space" for them to be creative. They saw Dr. Patel caring for them to the degree that she trusted them to write computer science programs "the way [they] desired," rather than chaining them to existing prototypes.

One male student specifically storied how he felt while enrolled in Professor Patel's course:

> I remember I was taking Organization – Computer Organization, and I was not the best test-taker, but, in that class, I remember I had a lot of fun learning. I think her extra credit assignment was actually to create a cyber Tic-Tac-Toe, and that took me a long time, but I remembered the process once it was done. I am, like, wow. I finally did it. I accomplished something, and I enjoyed that.

Another male mentioned that he had the opportunity to conduct research alongside Professor Patel. For that individual, exploring Educational Data Mining with Professor Patel was a "wonderful journey." Because of her caring approach, his research experience of data mining was "the best kind of help" he has ever received from a professor.

The Professors Build a STEM Community

Da Minh Nguyen, a senior Robert Noyce scholar majoring in mathematics in *teach*HOUSTON, a secondary STEM teacher education program, told us how the *teach*HOUSTON faculty as a whole strives to build a STEM community that made a difference in his learning, life, and future career trajectory. His experience with *teach*HOUSTON differed substantially from his previous learning experiences elsewhere on campus and in Vietnam, the country where he was born. In other classes he had taken, "people [were] divided into, like groups." In his view, the groups were based on "race or how long one has stayed in the United States." As a minority youth who is considered part of an Asian model majority (Craig, Zou, & Curtis, 2018), Da Minh often felt invisible in the presence of his professors and peers, which mirrors many other underrepresented minority students' experiences. Such was not the case with *teach*HOUSTON, he repeatedly impressed on us.

For Da Minh Nguyen, "*teach*HOUSTON is like a family," where the professors "held [his] hands" and "helped [him]." His master teacher, he emphasized, had been with him since he started the *teach*HOUSTON teacher education program. He especially learned from his master teacher not only "how teachers teach" but also how to professionally present himself in front of students, colleagues, administrators, and researchers in milieus such as classrooms, academic conferences, and NSF grant research group meetings. Also, as a Noyce scholar, he was able to participate in 300–400 hours of STEM research each summer. When a Stanford Research Lab opportunity opened up, he did not think he was ready for the opportunity. With the encouragement and help of his *teach*HOUSTON professors (particularly his master teacher), Da Minh applied for it and earned the chance to work with research scientists for three months in the Stanford lab. Da Minh concluded:

> I have learn[ed] a lot from them [*teach*HOUSTON professors] from a science perspective and also as a person. They taught me to believe in myself. And I learned that I am going to fail a lot, and I am going to learn a lot from failures.

The Professor Represents Females in the STEM Disciplines

Joyce Carr, a white female student in the *teach*HOUSTON Noyce Scholar Program, desired to be a teacher because of the support she received from her teachers when her family lived below the poverty line. Also, since childhood, she had been exposed to her parents' fundamentalist religious orientation, which mostly was expressed in her father's "archaic beliefs" (her description) about women. For example, he frequently commented that Joyce's mother "did not know anything because she came from Eve." Joyce naturally deduced that subtractive thinking about women was problematic because she had come to know – from her sustained interactions with her mother – was untrue. However, what Joyce learned in her family unit did not stop her from developing her critical thinking skills at school and feeding her burgeoning interest in the STEM disciplines, which historically have been dominated by males and reinforced by thinking somewhat akin to that of her father. At the core of her being, Joyce Carr knew from relating to other females and males (including her father) and as an inquirer herself that "mind has no sex" (De la Barre, 1673).

Unfortunately, the deficit thinking about females that Joyce first encountered in her family also occurred at university. A professor in one of Joyce's STEM classes "expect[ed] the males to do better in the class because males…have better mental manipulation of objects and more abstract concepts. And females… [are] just not good at doing that." Fortunately, because Joyce had previously heard comments like this, she was not as intimidated as she might have been by the professor's proclamation. She managed to tolerate the class because it was required, but she dropped another course after a similar experience. The young woman realized in a first-hand way that the dismissive behaviors of male professors could interfere with, and even deter, female students like her from studying the STEM disciplines:

> So, I felt – and that this could be why there are fewer women that go into STEM in general, because…that attitude seems to be quite prevalent. Oh, and in my other class with [another male] professor – I left after a week. I mean I…said something about a thermodynamic system, and he was very dismissive of me, like, no, no, no – even though what I said was overall right but there was a small flaw in my argument. Then some guy next to me yelled something out [that added to what I said]. The [male] professor then says, "That is it" – There is a problem with [differentiated treatment].

In this case, Joyce Carr transferred to Dr. Forrest's class and was able to flourish under Dr. Forrest's tutelage. In that class, she began to claim her right as a female to engage in scientific inquiry and to be a scientist. She explained:

> It [the experience] was –She is… the best [STEM] teacher on this campus. Dr. Forrest. And she is a woman, and she is incredible. And she did not disregard people's responses, did not shut people down whether male or female. I felt it was a much better environment, you know, and I did well in that class.

Joyce's decision to switch classes introduced her to a different STEM instruction. Dr. Forrest was supportive and motivating and did not adhere to "sexist stereotypes." Dr. Forrest's method of teaching helped Joyce to find her best-loved self (Craig, 2013; Schwab 1954/1978) as a female scientist and strengthened her preparation as a future STEM teaching professional.

The Professor Acts as a Navigator for Students

Numerous students complimented cybersecurity professor, Rish Gupta, for how he helped them to navigate their chosen areas of study and future careers. Dr. Gupta, a male of color himself, offered the course, Security Analytics, which is the first course on cybersecurity. Many students mentioned that the way Rish Gupta taught that course "opened up new worlds" for them. One African American female student had the following to say:

> Student: I loved it…It definitely is a big influence. Just seeing, you know, just even his knowledge of the area, it was just – to me it was amazing. It was exciting, of course, cause its cyber security, something I definitely want to go into. But just seeing, you know, just even the different methodologies within computer science and how they can be applied specifically to cyber security.
>
> Researcher: So, he really introduced you to that?
>
> Student: Yeah, he really did… He really exposed me to an exciting area of interest.

Another newly enrolled white male student mentioned that he did not plan to pursue a graduate degree in cybersecurity because he thought cybersecurity was mainly about attack and defense. However, Dr. Gupta's cybersecurity course led him to see how data mining, artificial intelligence, and many other fields relate to cybersecurity in highly engaging ways. It "triggered" his interest in further exploring the field. The same happened to an African American male. While enrolled in Rish Gupta's course, he awakened to the fact that cognitive science and cybersecurity could be combined in fruitful ways. This connection would be "unknown to [the student], had he not had lengthy discussions with Dr. Gupta."

A former Russian immigrant to the United States also talked about Rish Gupta as a thesis advisor who was in charge of his research project. According to this student, his advisor met with each student twice per week. "Dr. Gupta gives more homework than the rest," said the student. "It's not busy work," he explained, "You really earn your grades." For him, Rish Gupta is a "very reliable and very rigorous person." His Foundations of Security course has been especially useful to the student because the course relates to the area of the computer science industry where he is presently employed. He particularly credits his deep learning of "cross-eye scripting" and "proof-of-concept" to Rish Gupta's guidance and instruction.

A former Chinese immigrant student entered the program thinking that he had already found the area he wants to study for life. However, his interactions with Dr. Gupta led him to discover another area of specialization "in which [he] fell in love." This student first wanted to be a game developer as he was interested in graphics and specialization. However, his margins became stretched and his locus of possibilities widened after being introduced to the program and meeting Rish Gupta. The student is now specializing in human–machine generation. For this Chinese American student, Dr. Rish Gupta is an excellent role model from whom he "continues to learn."

Tom La, a Vietnamese immigrant student, also considered Dr. Gupta "a hero" in his eyes. When Tom was nearing graduation, Rish Gupta encouraged Tom to "look, look, look" for jobs that suited his interest and talents. He urged Tom to be persistent in his effort to achieve what Tom termed "the American Dream."

The majority of the computer science students also credited Professor Gupta with answering countless questions about their learning, research, and future careers. For them, he was/is their constant "go-to person." Rish Gupta "brokers teaching, research, and learning resources" in the cybertechnology and cyber-security fields. In short, he serves as a "navigator" (one student's expression) for his students where recruiting, advising, teaching, and career counseling is concerned.

The Professors Take an Inquiry Stance toward Teaching and Learning

In *teach*HOUSTON, "professors do not just talk about inquiry – they teach *using* inquiry and encourage us to learn that way, too." Almost every student we interviewed echoed this statement, uttered by a Latino male. They repeated it nearly every time we communicated with them in focus groups or via emails. The students collectively said (in the words of one white female) that they became keenly aware of "inquiry moves" and that their challenge "would not be how to use inquiry in the classroom," but to employ it "in school contexts where inquiry is not the norm and the [high stakes] accountability agenda rules instruction." In addition to these general statements, the *teach*HOUSTON students discussed three different ways inquiry used: (1) in a Physics by Inquiry course, (2) in learning about culturally relevant teaching pedagogies, and (3) in engineering education classes.

Ryan Spencer, a white male, entered university at 16 and dropped out for nearly 20 years soon after he claimed that "learning through inquiry" was the "missing piece" in his K-16 education. When he entered *teach*HOUSTON, he felt he had finally found his "educational home." He initially thought his white female physics education professor would teach Physics by Inquiry in a "simplistic way." However, she tasked groups of students to work on real-world problems with mirrors, pulleys, and other assorted hands-on materials. Her approach caused him to see how "rigorous and relevant" teaching and learning through inquiry could be – not only for himself as a teacher but also for himself as a "learner amid a classroom of learners."

Another student – a Latino female – specifically spoke about learning from her African American female professor how to use inquiry to teach in a culturally relevant way. The female student, like many of her peers, saw, through carefully crafted examples, how different teaching strategies and use of different materials could open up productive discussions about human differences. This female student specifically remembered classroom discussions about "scientists historically being male and mostly white." The conversation that ensued awakened the Latino female to the legacy of STEM and how she might introduce "other stories of other people" to future discussions taking place in her high school classroom.

A female engineering education professor also influenced a bevy of *teach*HOUSTON students through her use of inquiry to teach engineering as a field of study. Interestingly, students lauded different aspects of this particular professor's pedagogy. One Vietnamese female was highly influenced by the Middle Eastern American professor's manner and presence. The professionalism the STEM professor exuded was not lost on the Vietnamese landed immigrant. For another white female, it was the professor's way of rolling out courses – how pieces of the inquiry puzzle seemed to magically appear – and fit together like "hand in glove." In addition, for other several males and females of all races, it was how this professor paced her questions in order to deepen conceptual learning. They thought it was "masterful" how she "effortlessly accomplished these complex tasks – and all at once."

While we have not discussed the inquiry moves of every *teach*HOUSTON faculty member, this sampling of three diverse professors provides readers with a flavor of how inquiry became lived by students in particular NSF-funded programs.

INTERPRETIVE THEMES

When our author team traversed our five narrative exemplars, we identified three overarching themes common to them: (1) agency as a student and agency as a human being, (2) development of students' multilayered identities, and (3) professors' engagement of themselves in their interactions with students.

Agency as a Student and Agency as a Human Being

In the five narrative exemplars that we featured, students' agency in learning, with the support from their professors, was enhanced and intertwined with their agency as human beings. Dr. Patel's extra credit assignments offered students a space to actively explore their field of study in a fun way. During the process, the students not only became agentive learners but also human beings who could think creatively, imaginatively, and critically about nearly any topic of investigation. In the community built by his *teach*HOUSTON professors, Da Minh Nguyen's growth was relationally, emotionally, and intellectually evident. His graphic narration showed him asserting his opinion and more publicly revealing his sense of self – even in focus group situations where others held contrary views. Additionally, Da Minh, with a strengthened belief in himself, started pursuing his personal and professional goals without fear of failure.

*teach*HOUSTON student, Joyce Carr, having learned from physics professor, Dr. Forrest, self-authored (Nadelson et al., 2017) her personal position concerning gender issues. Other students, guided by Rish Gupta, expanded their horizons of knowing and found new areas of interest in the cybersecurity field, a field that was almost entirely new to them. In different ways, the students were facilitated to "actively navigate the difficult terrain of the construction of personal meaning" (Paris & Combs, 2006, p. 573) in becoming STEM professionals. In sum, the professors helped shift students' views of teaching, learning, and human differences through active, empathetic, practically oriented scientific inquiry.

The students' narratives further illuminated how their agency as learners and their agency as human beings were nourished by their professors. The professors did not consider course content as an end in itself, but employed content as a way to facilitate students to learn how to learn (Weimer, 2002). Moreover, the professors helped students to realize that "nothing in life is won without discipline, hanging in, being able to last through periods of mere slogging until something greater grows out of them" (Taylor, 1985a, p. 27). Thus, students were able to experience learning in meaningful ways, rather than being occupied with reproducing information delivered by the professors (Wright, 2011). Each of the students was not prepared to be a professional with "just hands or a brain," but as a "whole person" (De Weerdt, Bouwen, Corthouts, & Martens, 2006, p. 324) directed by their "reflective, integrated agency" (Mackenzie, 2008, p. 9), persistent in achieving their goals and courageous in handling uncertainties in their life and career trajectories.

Development of Students' Multilayered Identities

The development of students' multilayered identities is evident throughout their interactions with professors. First, students' identities as active learners were cultivated by and validated by their professors. Because university students are presumably equipped with independent learning abilities, professors often emphasize subject matter knowledge in their teaching, believing students would naturally process what is taught. Thus, student learner identities become reduced to passive receivers of knowledge. However, the exemplars in our study presented quite a different picture. For instance, the student who considered himself not good at taking tests found Dr. Patel's optional tasks valuable opportunities for him to recover his identity as a capable and creative learner. Also, Ryan Spencer, who dropped out from university for two decades, discovered "learning through inquiry" as the "missing piece" that he needed to repair his impoverished identity (Lindemann Nelson, 2001) as a learner. Even Joyce Carr, who found her voice in high school, still needed a space to focus solely on knowing and doing science rather than constantly having to prove that she, as a female, belonged in the program and that she, too, could think like a scientist and enter a STEM career.

Although cultivating the learner identity of students was important, it was not the sole focus of the professors in our study. The professors we spotlighted awakened the multiple facets of students' identities through calling students' attention to "otherness" (Latta & Kim, 2009, p. 141) in verbal and nonverbal ways. The *teach*HOUSTON professors treated Da Minh as a preaccredited (De Weerdt et al., 2006) professional and showed him how to interact with different people in various contexts while maintaining a firm belief in himself. Under the influences of inquiry teaching, the Latino female student enrolled in the *teach*HOUSTON program solidified her identity as a teacher who would value "other stories of other people" in order to promote students' understanding of human differences. Dr. Rish Gupta helped his students to see rich possibilities inherent in integrating other disciplines with cybersecurity, which enhanced students' dreams of who they could be in their STEM careers. He, along with other professors already mentioned, illuminated how "otherness [is] intrinsically linked within the unfolding of [students'] unknown sel[ves] in the present and future" (Latta & Kim, 2009, p. 143).

In sum, student identity, "rather than being fixed and predefined, is perpetually emergent, becoming increasingly multiplex" (Greene, 1988, p. 23) as knowledge gleaned from experience becomes wider and deeper. Acting as guides on the side (Weimer, 2002), professors like Dr. Patel, Dr. Forrest, Dr. Gupta, and the *teach*HOUSTON faculty provided students with more perspectives to gear their learning, career, and lives in the direction that follows their own authentic ways of being (Taylor, 1991/2003). Therefore, students became not only "self-sufficient, independent, and creative thinkers who appreciate and value the subject [that they learned]" (Brown, 2008, p. 33) but also "individual human beings able to make choices and act on these choices in a way that makes a difference in their lives" (Martin, 2004, p. 135).

Professors' Engagement of Themselves in Their Interactions with Students

While it is commonly agreed that student engagement plays a significant role in students' learning and identity formation, the existing national and international research mainly focuses on strategies faculties employ to increase student engagement (Quaye, Harper, & Pendakur, 2019). However, the professors' praxis in our narrative exemplars revealed that increasing student engagement was much more than a technical issue to resolve. Professors' implicit and explicit engagement of their diverse selves in their interactions with students was exceedingly important as well. In short, the teaching role and teaching strategies on their own did not make the difference for the undergraduate and graduate students. Rather, it was the person who each professor was shaped how the students were taught and how and what students learned (Hansen, 1995).

For example, Dr. Patel showed deep interest and cared about students' ideas. She intentionally fostered positive attitudes toward the creative process and nurtured their careers in view. As for the *teach*HOUSTON professors, they treated Da Minh as a family member, helping him to feel he belonged in the program, walking him through situations of uncertainty and ambiguity (Baxter Magolda, 2007). Further to this, underrepresented minority professors (i.e., *teach*HOUSTON, physics, and cybertechnology) modeled their personal professional beings on an ongoing basis and guided the featured students – often cross culturally – to discover what they believed and valued as future STEM professionals and citizens. Moreover, when students stood at decision-making crossroads "not knowing which road to take" in the future, Dr. Gupta closely interacted with students to uncover students' true needs, which further "enlarge[d]" and challenge[d] the [students'] understandings" (Latta & Kim, 2009, p. 143) of who they could be.

Thus, we see how the featured professors in this chapter, through involving themselves in their life's work, facilitated students to build or repair relationships between themselves and STEM subject matter/STEM learning (i.e., inquiry)/ STEM career/life. As mentioned, "identity is a moving intersection of the inner and outer forces that make [people] who [they are] …" (Palmer, 1998, p. 13). The self-infused praxis of the professors awakened students to their capacities to learn both intellectually and emotionally. Their approach activated students' inner forces through connecting "externally validated knowledge" to students' "inner ways of knowing" (Hunter, Laursen, & Seymour, 2007, p. 67). Although the six NSF grants mainly aimed to prepare students as qualified STEM workers, we see how the professors went far beyond this to help students flourish as self-fulfilled human beings with rich professional identities and hopes and dreams for which to strive.

CLOSING STATEMENTS

In this chapter, students' narratives in the form of exemplars caused student-centeredness and professors' professional development in higher education to bubble to the fore as particularly important considerations. We quickly recognized that simply talking about individuals or "doing education to and for learners" (McCombs & Whisler, 1997, p. xi) does not guarantee student-centered teaching and learning in classrooms. It is the students' experience which gives "lived meanings" (Paris & Combs, 2006, p. 574) to learner-centeredness. As illustrated in this chapter, professors "*look with* the learner at what learning means and how it can be enhanced from within – drawing on the learner's unique perspectives, talents, capacities, and experiences" (McCombs & Whisler, 1997, p. xi, italics in original). Through this process, the "unique, personal humanness of meaning making" (Latta & Kim, 2009, p. 143) inherent in professor–student interactions assists both sides in unraveling "relational complexities" (p. 141) in diverse contexts which shape and are shaped by rapidly changing STEM fields and STEM professions.

Conventionally, subject matter knowledge (Gaff & Simpson, 1994) and pedagogy (Wieman, 2017) have been the central focus of developing professors as professionals. It is also not uncommon that universities make evidence-based and research-based approaches available to faculty in the form of promising best practices (Oleson & Hora, 2014), which may be disconnected from one another and from individual professor's inclinations. Those students' storied experiences implied potential directions that professor professional development might take in the future.

At first, our study illustrated the necessity of professors building and capitalizing on "connectedness and relevancy" (Latta & Kim, 2009, p. 140) from student, subject matter, learning, career, and life perspectives. This helps students to grow as life-long learners and to adjust and refresh their knowledge and skills on the rapidly evolving STEM landscape. Second, we know as well from the narrative exemplars that the college students gravitated toward professors who had solid personal and professional identities, albeit still in the making. What we learn here is that an overemphasis on knowledge and research may distance professors from teaching, students, and themselves as human beings playing public roles. The dynamic interplay between professors' personal and professional selves presents critically important moments for developing students' "sense of self in both the public and private [spheres]" (Peel, 2005, p. 496). Thus, professors with their integrated beings can "elicit the significance of the students" (Latta & Kim, 2009, p. 141), which helps students to unearth additional opportunities for growth. Along the way, the professors loosened their subject matter straitjackets and played a more organic, interactive role for students. Each of the professors thus became a "mentor, guide, model; ally of the student against ignorance, participant with the student in high adventures into the worlds of intellect and sensibility" (Schwab, 1969, p. 20).

We, as authors, sincerely hope that our narrative inquiry into the influence of professors on students entering the STEM fields and careers "incite[s] a turn and re-turn to [readers'] self-understandings" (Latta & Kim, 2009, p. 144) of their own experiences. Through "acting on possibilities again and again" (p. 144) via an "ever-present process" (Dewey, 1938, p. 50) of inquiry, the influence of professors will spread from student to student and milieu to milieu (university and workplace) locally, nationally, and internationally.

AUTHORS' NOTE

This research was funded by the following NSF grants:

- Noyce Grant: *Recruitment, Preparation and Retention of STEM Students as High School Teachers* [1240083]
- S-STEM Scholar Program: *Preparing Computer Science Students for Global Challenges of the twenty-first Century* [1356705]
- CyberCorps (R): *Scholarship For Service (SFS): Increasing Talented Trusted Professionals* [1433817]
- Noyce Grant Track 1: *STEM Teacher Preparation at the University of Houston: Learning through Informal and Formal Experiences* (UH-LIFE) [1557309]
- STEM Scholarship Program: *Promotion and Retention of STEM Education through Networking Team* (PARENT) Support [1644191]
- Noyce Grant Track 3: *Leadership through Advocacy and Development in STEM* (UH-LEAD) [1759454]

REFERENCES

Bamberg, M., & Georgakopoulou, A. (2008). Small stories as a new perspective in narrative and identity analysis. *Text & Talk*, *28*(3), 377–396.

Baxter Magolda, M. B. (2007). Self-authorship: The foundation for twenty-first-century education. *New Directions for Teaching and Learning*, *2007*(109), 69–83.

Beauchamp, C., & Thomas, L. (2009). Understanding teacher identity: An overview of issues in the literature and implications for teacher education. *Cambridge Journal of Education*, *39*(2), 175–189.

Beijaard, D., Meijer, P., & Verloop, N. (2004). Reconsidering research on teachers' professional identity. *Teaching and Teacher Education*, *20*(2), 107–128.

Bonous-Hammarth, M. (2000). Pathways to success: Affirming opportunities for science, mathematics, and engineering majors. The *Journal of Negro Education*, *69*(1–2), 92–111.

Broecke, S., & Hamed, J. (2008). *Gender gaps in higher education participation: An analysis of the relationship between prior attainment and young participation by gender, socio-economic class and ethnicity*. DIUS Research Report 08-14. Retrieved from https://dera.ioe.ac.uk/8717/1/DIUS-RR-08-14.pdf.

Brown, J. K. (2008). Student-centered instruction: Involving students in their own education. *Music Educators Journal*, *94*(5), 30–35.

Chang, M. J., Eagan, M. K., Lin, M. H., & Hurtado, S. (2011). Considering the impact of racial stigmas and science identity: Persistence among biomedical and behavioral science aspirants. *The Journal of Higher Education*, *82*(5), 564–596.

Clandinin, D. J., & Connelly, F. M. (2000). *Narrative inquiry: Experience and story in qualitative research*. San Francisco, CA: Jossey-Bass.

Clandinin, D. J., & Rosiek, J. (2007). Mapping a landscape of narrative inquiry: Borderland spaces and tensions. In D. J. Clandinin (Ed.), *Handbook of narrative inquiry: Mapping a methodology* (pp. 35–75). Thousand Oaks, CA: Sage Publishing.

Connelly, F. M., & Clandinin, D. J. (1990). Stories of experience and narrative inquiry. *Educational Researcher*, *19*(5), 2–14.

Cotton, S. R., & Wilson, B. (2006). Student-faculty interactions: Dynamics and determinants. *Higher Education*, *51*, 487–519.

Craig, C. (2013). Teacher education and the best-loved self. *Asia Pacific Journal of Education*, *33*(3), 261–272.

Craig, C., Evans, P., Verma, R., Stokes, D., & Li, J. (2018). A tribute to "unsung teachers": Teachers' influences on students enrolling in STEM programs with the intent of entering STEM careers. *European Journal of Teacher Education*, *42*(3), 335–358.

Craig, C., Zou, Y., & Curtis, G. (2018). The developing knowledge and identity of an Asian-American teacher: The influence of a China study abroad experience. *Learning, Culture and Social Interaction*, *17*, 1–20.

De Weerdt, S., Bouwen, R., Corthouts, F., & Martens, H. (2006). Identity transformation as an intercontextual process. *Industry and Higher Education*, *2*(5), 317–326.

De la Barre, F. (1673). On the equality of the two sexes: A physical and moral discourse, which shows that it is important to rid oneself of prejudice. Retrieved from https://plato.stanford.edu/archives/win2019/francois-barre/.

Dewey, J. (1938). *Experience and education*. New York, NY: Basic Books.

Dortch, D., & Patel, C. (2017). Black undergraduate women and their sense of belonging in STEM at predominantly White institutions. *NASPA Journal About Women in Higher Education*, *10*(2), 202–215.

Freeman, K. E., Alston, S. T., & Winborne, D. G. (2008). Do learning communities enhance the quality of students' learning and motivation in STEM? *The Journal of Negro Education*, *77*(3), 227–240.

Gaff, J. G., & Simpson, R. D. (1994). Faculty development in the United States. *Innovative Higher Education*, *18*(3), 167–176.

Giffi, C., Dollar, B., Drew, M., McNelly, J., Carrick, G., & Gangula, B. (2015). The skills gap in U.S. manufacturing 2015 and beyond. Retrieved from http://www2.deloitte.com/content/dam/Deloitte/us/Documents/manufacturing/us-pip-the-manufacturing-ins titute-and-deloitte-skills-gap-in-manufacturing-study.pdf.

Greene, M. (1988). *The dialectic of freedom*. New York, NY: Teachers College Press.
Greene, M. (1995). *Releasing the imagination: Essays on education, the arts, and social change*. San Francisco, CA: Jossey-Bass.
Hansen, D. (1995). *The call to teach*. New York, NY: Teachers College Press.
Heikkinen, H. L., Huttunen, R., & Syrjälä, L. (2007). Action research as narrative: Five principles for validation. *Educational Action Research*, *15*(1), 5–19.
Heikkinen, H. L., Huttunen, R., Syrjälä, L., & Pesonen, J. (2012). Action research and narrative inquiry: Five principles for validation revisited. *Educational Action Research*, *20*(1), 5–21.
Hunter, A., Laursen, S. L., & Seymour, E. (2007). Becoming a scientist: The role of undergraduate research in students' cognitive, personal, and professional development. *Science Education*, *91*(1), 36–74.
Johnson, A. (2007). Unintended consequences: How science professors discourage women of color. *Science Education*, *91*(5), 805–821.
Johnson, D. R. (2007). *Sense of belonging among women of color in science, technology, engineering, and math majors: Investigating the contributions of campus racial climate perceptions and other college environments*. Doctoral Dissertation, University of Maryland.
Kennett, J., & Matthews, S. (2008). Normative agency. In C. Mackenzie, K., & Atkins (Eds.), *Practical identity and narrative agency* (pp. 212–230). New York, NY: Routledge.
King, T. (2003). *The truth about stories*. Toronto, CA: Anansi Press.
Latta, M. M., & Kim, J. H. (2009). Narrative inquiry invites professional development: Educators claim the creative space of praxis. *The Journal of Educational Research*, *103*(2), 137–148.
Lincoln, Y., & Guba, E. (1985). *Naturalistic inquiry*. Beverly Hills, CA: Sage Publishing.
Lindemann Nelson, H. (2001). *Damaged identities: Narrative repair*. Ithaca, NY: Cornell University Press.
Lyons, N., & LaBoskey, V. (2002). *Narrative inquiry in practice: Advancing the knowledge of teaching*. New York, NY: Teachers College Press.
Mackenzie, C. (2008). Introduction: Practical identity and narrative agency. In C. Mackenzie & K. Atkins (Eds.), *Practical identity and narrative agency* (pp. 1–28). New York, NY: Routledge.
Malone, K. R., & Barabino, G. (2009). Narrations of race in STEM research settings: Identity formation and its discontents. *Science Education*, *93*(3), 485–510.
Marginson, S., Tytler, R., Freeman, B., & Roberts, K. (2013). *STEM: Country comparisons: International comparisons of science, technology, engineering and mathematics (STEM) education*. Final report. Australian Council of Learned Academies, Melbourne, Vic.
Marshall, M. N. (1996). Sampling for qualitative research. *Family Practice*, *13*(6), 522–526.
Martin, J. (2004). Self-regulated learning, social cognitive theory, and agency. *Educational Psychologist*, *39*(2), 135–145.
McCombs, B., & Whisler, J. (1997). *The learner-centered classroom and school: Strategies for increasing student motivation and achievement*. San Francisco, CA: Jossey-Bass.
Ministry of Education of China. (1993). Jiaoyu Gaige he Fazhan Gangyao (Outline for education reform and development). Ministry of Education of China, Beijing.
Mishler, E. (1990). Validation in inquiry-guided research: The role of exemplars in narrative studies. *Harvard Educational Review*, *60*(4), 415–443.
Nadelson, L. S., McGuire, S. P., Davis, K. A., Farid, A., Hardy, K. K., Hsu, Y. C., & Wang, S. (2017). Am I a STEM professional? Documenting STEM student professional identity development. *Studies in Higher Education*, *42*(4), 701–720.
Oleson, A., & Hora, M. T. (2014). Teaching the way they were taught? Revisiting the sources of teaching knowledge and the role of prior experience in shaping faculty teaching practices. *Higher Education*, *68*(1), 29–45.
Palmer, P. (1998). *The courage to teach: Exploring the inner landscape of a teacher's life*. Hoboken, NJ: John Wiley & Sons.
Paris, C., & Combs, B. (2006). Lived meanings: What teachers mean when they say they are learner-centered. *Teachers and Teaching: Theory and Practice*, *12*(5), 571–592.
PCAST. (2012). *Engage to excel: Producing one million additional college graduates with degrees in science, technology, engineering, and mathematics*. Washington, DC: PCAST.

Peel, D. (2005). Peer observation as a transformatory tool? *Teaching in Higher Education, 10*(4), 489–504.

Pollard, A. (Ed.). (2010). *Professionalism and pedagogy: A contemporary opportunity. A commentary by the teaching and learning research programme and the general teaching council for England.* London: TLRP.

Quaye, S. J., Harper, S. R., & Pendakur, S. L. (Eds.). (2019). *Student engagement in higher education: Theoretical perspectives and practical approaches for diverse populations.* Abingdon: Taylor & Francis.

Ricoeur, P. (1988). *Time and narrative* (Vol. III). K. McLaughlin & D. Pellauer (Trans.). Chicago, IL: University of Chicago Press.

Ricoeur, P. (2003a). Life in quest of narrative. In D. Wood (Ed.), *On Paul Ricoeur: Narrative and interpretation* (pp. 20–33). London: Taylor & Francis.

Ricoeur, P. (2003b). Narrative identity. In D. Wood (Ed.), *On Paul Ricoeur: Narrative and interpretation* (pp. 188–199). London: Taylor & Francis.

Schleicher, A. (2018). *Valuing our teachers and raising their status: How communities can help.* Paris: OECD.

Schwab, J. (1954/1978). Eros and education: A discussion of one aspect of discussion. In I. Westbury & N. Wilkof (Eds.), *Science, curriculum and liberal education: Selected essays.* Chicago, IL: University of Chicago Press.

Schwab, J. (1964). Structure of the disciplines: Meanings and significances. In G. Ford, & L. Pugno (Eds.), *The structure of knowledge and the curriculum.* Chicago, IL: Rand McNally & Company.

Schwab, J. (1969). The practical: A language for curriculum. *The School Review, 78*, 1–23.

Sithole, A., Chiyaka, E. T., McCarthy, P., Mupinga, D. M., Bucklein, B. K., & Kibirige, J. (2017). Student attraction, persistence and retention in STEM programs: Successes and continuing challenges. *Higher Education Studies, 7*(1), 46–59.

Smeby, J. (2000). Same-gender relationships in graduate supervision. *Higher Education, 40*, 53–67.

Smith, C. L., & Allen, J. M. (2006). Essential functions of academic advising: What students want and get. *Nacada Journal, 26*(1), 56–66.

Taylor, C. (1985a). The concept of a person. In C. Taylor (Ed.), *Human agency and language: Philosophical papers 1* (pp. 97–114). New York, NY: Cambridge University Press.

Taylor, C. (1985b). What is human agency? In C. Taylor (Ed.), *Human agency and language: Philosophical papers 1* (pp. 15–44). New York, NY: Cambridge University Press.

Taylor, C. (1985c). Self-interpreting animals. In C. Taylor (Ed.), *Human agency and language: Philosophical papers 1* (pp. 45–76). New York, NY: Cambridge University Press.

Taylor, C. (1991/2003). The sources of authenticity. In C.Taylor (Eds.), *The ethics of authenticity*, pp. 25–30). Cambridge, MA: Harvard University Press.

Trolian, T. L., & Parker III, E. T. (2017). Moderating influences of student–faculty interactions on students' graduate and professional school aspirations. *Journal of College Student Development, 58*(8), 1261–1267.

Weimer, M. (2002). *Learner-centered teaching: Five key changes to practice.* San Francisco, CA: Jossey-Bass.

Wieman, C. (2017). *Improving how universities teach science.* Harvard, MA: Harvard University Press.

Wright, G. B. (2011). Student-centered learning in higher education. *International Journal of Teaching and Learning in Higher Education, 23*(1), 92–97.

Xu, Y. J. (2016). Attention to retention: Exploring and addressing the needs of college students in STEM majors. *Journal of Education and Training Studies, 4*(2), 67–76.

THE VALUE OF STEM SCHOLARSHIP GRANTS TO UNDERGRADUATE AND GRADUATE STUDENTS INTENDING TO STUDY THE STEM DISCIPLINES AND PURSUE STEM CAREERS

Jing Li, Cheryl J. Craig, Tenesha Gale, Michele Norton, Gang Zhu, Paige K. Evans, Donna W. Stokes and Rakesh Verma

ABSTRACT

This chapter narratively examines the value of scholarship grants to seven underrepresented science, technology, engineering, and mathematics (STEM) students who attended the same research comprehensive university. The scholarships that the students in our convenience sample received were awarded by six National Science Foundation grant programs in the United States. A literature review tracing the effects of scholarships, instrumentalism, and the core purpose of education sets the context for this narrative investigation. The four pillars comprising the theoretical framework are value, experience, story, and identity. The seven stories of impact that emerged from the narrative inquiry reveal multiperspectival insights into the value of scholarships to students' lives, careers, and selves. Moreover, we also explore how scholarship recipients established their sense of value in autonomous and committed ways while promoting their personal welfare and seeking the common good of others. All of these important considerations contribute to the national and

Preparing Teachers to Teach the STEM Disciplines in America's Urban Schools
Advances in Research on Teaching, Volume 35, 179–200

ISSN: 1479-3687/doi:10.1108/S1479-368720210000035011

international literature relating to diversity, higher education, STEM careers, and the power of scholarship grants to transcend instrumentalism privileging workforce demands.

Keywords: STEM; scholarship grants; underrepresented students; higher education; narrative inquiry; STEM careers; *teach*HOUSTON

In the twenty-first century, the development of science, technology, engineering, and mathematics (STEM) fields plays transformative roles in social change (Christensen, Knezek, & Tyler-Wood, 2014; Miller, Vaccaro, Kimball, & Forester, 2020; Wang, Ye, & Degol, 2017). However, many nations have forecasted shortfalls in their STEM workforce (e.g., Marginson, Tytler, Freeman, & Roberts, 2013; Smith, 2011). In 2012, the President's Council of Advisors on Science and Technology (PCAST) in the United States predicted that the American STEM workforce would soon encounter a deficit of one million college graduates. Therefore, an urgent need exists for higher education to increase the recruitment and retention of college students who study the STEM disciplines with the intent of pursuing STEM careers (Diekman & Benson-Greenwald, 2018).

In the past decade, women have outnumbered men in college enrollment and minority students have enrolled in college at record numbers (Smith, 2009; Worthington, Hart, & Khairallah, 2010). The rapid shifts in demographics in higher education present both opportunities and challenges for the task of expanding the STEM talent pool. On one side, there are potentially more unrepresented students who may choose and sustain in the STEM fields. On the other side, the exit rates have been especially high for the "underrepresented majority" (women and racial/ethnic minorities) enrolled in STEM majors (Riegle-Crumb, King, Grodsky, & Muller, 2012; Seymour & Hewitt, 1997; Smith, 2011), although women and minority participation rates in STEM have seen modest increases since 1989 (National Science Foundation, 2013). For instance, even though Latinos who earned bachelor's degrees in physics, engineering, mathematics, and statistics increased from 2001 to 2010, they remained "woefully underrepresented" (Malcom & Malcom-Piqueux, 2013, p. 177) compared to the whole college population. Additionally, African American participation in the same majors also declined from 2001 to 2010.

Many studies have investigated the attrition of underrepresented students in STEM disciplines. It was found that inadequate academic preparation, lack of role models/mentors, marginalization, skepticism, self-doubt, and isolation (Chang, Eagan, Lin, & Hurtado, 2011; Dortch & Patel, 2017; Harris, 2015; Rodriguez & Blaney, 2020) are challenges those students frequently encounter in their college journey and may cause them to leave the STEM majors. Frequently employed strategies to sustain minority students include providing supportive institutional environment and organizing friendly and collaborative peer groups (Clewell & Ficklen, 1987; Museus, Palmer, Davis, & Maramba, 2011; Ost, 2010; Stout, Grunberg, & Ito, 2016).

Another strategy is to provide financial incentives for STEM students because many of them, especially the underrepresented students, are often limited by the availability of sufficient finances with regard to their college experiences (Graham, 2017). The six National Science Foundation (NSF)–funded scholarship programs in our study are exemplars of this theory of action. They provide scholarships which subsidize or pay for university students' education for a designated period of time, sometimes in exchange for work in a STEM-related field. Quite a few studies have explored how financial incentives relate to students' retention in the STEM disciplines and professions (e.g., Hungerford & Hodgson, 2013; Masterton, Moss, Korin, & Watters, 2014; Schuurs & Amsters, 2010). Although the above correlations are an important area of investigation, sustaining qualified STEM students is more than persistence to graduation and career aspirations (Schreiner, 2010). The value of scholarship grants – beyond preparing students for future STEM careers from knowledge and skill perspectives – has largely not been examined. Therefore, a deep understanding of the possible shaping influences of scholarship-bearing programs on students is needed, which is the research gap our study bridges.

Our study centers around the personal and learning experiences of seven student participants who are enrolled in one or two of the six NSF STEM scholarship programs. The seven students are all underrepresented students (women or immigrant students) in the STEM fields. From students' perspectives, our study presents the "natural joints" (Schwab, 1980, p. 365) each individual student has with their scholarship-bearing programs. Put differently, we inquire into how the scholarship programs intertwine with students' personal experiences in multiplicity of ways, which further affect students' learning, career paths, well-being, and their unfolding sense of becoming. All those combine to contribute to the quality of underrepresented students' personal and professional lives. This knowledge is vital to NSF, universities, and other national and international agencies, who sponsor the sustaining of unrepresented students in the STEM disciplines. This also provides insights for fostering diversity in higher education.

NSF SCHOLARSHIPS IN STUDY

In 1950, the NSF was founded with a mission to promote the development of science, health, prosperity, and welfare of the nation. Since then, NSF has funded research and development in the STEM disciplines in order to prepare more university students to fill the STEM workforce needs in the future.

The six NSF grants in our study include one Scholarship for Service (SFS) program, two S-STEM grant programs, and three NOYCE program awards. All six grants are awarded to the University of Houston, the second most diverse, research-intensive, Asian-serving, Hispanic-serving campus in the United States. The SFS program in the Department of Computer Science offers graduate students a full scholarship for education and training, with the intent of preparing superior cybersecurity professionals who must work after graduation for government agencies for a period equal to the length of their scholarship awards. The two S-STEM grants are located respectively in the Department of Computer Science and Department of Physics. Through multidisciplinary approaches, they prepared STEM students who can meet the future workforce needs and contribute to advancing America's economic prosperity. Sponsored by three Noyce program awards, *teach*HOUSTON is an innovative approach to STEM teacher education. This program is a collaboration between College of Natural Science and Mathematics, College of Education, and urban school districts. *teach*HOUSTON prepares undergraduate and graduate students majoring in mathematics and/or science to become qualified STEM high school teachers who will inspire the next generations of students to discover, create, and lead. Much like the SFS cybersecurity grant program, graduated Noyce scholars must serve in high-needs urban schools for a period of time that depends on the number of semesters they receive the scholarship. Part of the agenda in all six NSF STEM scholarship programs is to recruit and sustain more students of diversity in their respective fields.

LITERATURE REVIEW

We start this section with a review of current research on the effects of scholarships on students. The literature's seeming emphasis on the correlations between financial incentives, students' academic achievement, and career trajectory led us to dive into two other research strands: instrumentalism and the purpose of education. The intertwined strands lay a foundation for us to view the influences of scholarships on students from a more enriched perspective.

Effects of Scholarships

As foreshadowed, financial incentives/scholarships have been a widely used strategy to support student development in higher education. This is especially the case for promising students of color who come from low-income families and communities which may have been historically underserved. This is also true of students of all races who are the first generation in their families to attend colleges. Furthermore, according to Fenske and Associates (2000), "The National Action Council for Minorities in Engineering states the availability of adequate financial aid resources is among the top five factors related to minority persistence in Engineering" (p. 69). Buykx, Humphreys, Wakerman, and Pashen (2010), among other workforce maintenance strategies, also found that financial incentives help retain workers in jobs that are traditionally difficult from a recruitment perspective.

Liou and his colleagues conducted studies on the influences of the Noyce Scholarship Program on students. They (2010) found that students mainly perceived the Noyce program "as a means to complete their teacher education program; as a means to become teachers in high-need schools" (p. 451). Thus, how the program impacted the Noyce Scholar's commitment to teaching seemed to be an aspect less important to the scholarship recipients. In a national study with principal investigators of the Noyce programs, Lawrenz and his colleagues (2008) found that 88 percent of them believed that Noyce funding had a lesser effect on enrolled students' perceptions about teaching careers, relationships with the surrounding community, school districts, industry, and STEM faculty at their institutions. Liou and Lawrenz (2010) further argued that variables such as Noyce scholars' race, their pathway into teaching, and the amount of funding they received could impact recipients' perceptions of scholarship influences. This indicated that there are more complicated factors and dynamics to know about in order to fully encompass and understand the influences of scholarship grant programs on students.

In sum, when the effects of scholarships are discussed, the education research, national organizations, and scholarship recipients have a tendency to mainly focus on students' recruitment, academic achievement, and staying in the STEM profession. The driving force behind this orientation is the means-ends reasoning of instrumentalism, which will be discussed in the next session.

Instrumentalism

Pinar (2012) metaphorically described education through the instrumentalist lens of "a complex automobile engine. If only [people] make the right adjustments in teaching, in learning, in assessment – it will hum and transport people to their destination…" (p. 170). This means-ends, instrumentalist reasoning sits in the backdrop of many educational policies in many countries. In the United States, when the government's goals of ensuring the preeminence of the country's economy became projected in the field of education, NSF issued calls to attract, recruit, train, and sustain more qualified STEM students. Hence, the STEM scholarship programs we introduced earlier are examples of "means" to achieve national goals, which are intended "ends."

John Dewey (1916), however, warned that externally imposed ends might "limit intelligence…leav[ing] to the latter [intelligence] nothing but a mechanical choice of means" (p. 111). Thus, if meeting the national workforce needs becomes the single focus of the STEM scholarship programs, the programs would risk prioritizing strategies of preparing students as professional technicians, rather than other aspects which help to cultivate students as integrated human beings (Palmer, 1998). As a result, students, teachers, programs, and universities may gradually lose an appreciation (Phelan, 2009) for "the distinction between utility and meaningfulness...between in order to and for the sake of" (Arendt, 1998, p. 157). The STEM scholarship programs would risk being "reduc[ed]... to anthropocentric projects of calculation, strategy, and human mastery" (Phelan, 2009, p. 107). Inquiries of multiperspectival "why[s]" are easily usurped by the single question of how (Arendt, 1998) to reach the predetermined goals. What would be ignored is "the manner in which schools [universities] shape and are shaped by social inequities" and "the complex responsibility of...educators toward the life of [students] and for the continuance of the world" (Phelan, 2009, p. 107), not to mention the deliberation about the purpose of education – a core element which influences education practices with regard to what is taught, how to teach, and in what ways teachers and students are evaluated.

The Purpose of Education

The purpose of education interfaces with many factors such as a nation's politics, economy, and culture, as shown in the goals of NSF scholarship programs in our study. Contentious debates about the purpose of education have occurred throughout history (i.e., Socrates vs Athens). In the United States, presidential rhetoric is an important source (Carpenter, 2005) for understanding the changing dynamics in the nation's purpose of education, values, and other trends. Through an analysis of inaugural and State of the Union speeches (1790–2002), Carpenter (2005) found that civic responsibility and economic efficiency are two purposes that have overwhelmingly been addressed through education, with the emphasis shifting from the former to the latter over time. Giroux (1988) also observed that "education reform has become synonymous with turning schools into 'company stores' and defining school life primarily in terms that measure their utility against their contribution to economic growth and cultural uniformity" (p. 18).

Meanwhile, based on Carpenter's (2005) analysis, the other main categories – self-realization and human relationship – have received scant attention. Themes such as the individual's development of self, individual's place in the community (Beane, 1998), and "the tensions between individual freedom and social equality" (Carpenter, 2005, p. 283) are only discussed in abstract and general terms. However, the students, such as the participants in our study, came to education with "intrinsic activities and needs" (Dewey, 1916, p. 114). Although the goal of becoming a qualified STEM professional is what they expect/are expected to achieve, they are not simply being satisfied as "end users of codified knowledge," which is "abstracted from context, extracted from persons, and devoid of relationship" (Craig, 2020). Rather, students actively make a "finely grounded sense" of their "life personal meaning[s]" (Goodson, 2007, p. 133) in their education experiences. Students' selves thus evolve and "equip [students] to function socially, politically," (Carpenter, 2005, p. 283) and professionally, which in turn helps them to build a greater sense of self (Hartoonian, 1999). Our study intends to present empirical evidence of this significant circular relationship that easily goes unnoticed in educational research. Put differently, our study illuminates how students "[reconstruct] private meaning and public purpose" (Goodson, 2007, p. 147) of the scholarship grants to live and relive their stories of experience.

CONCEPTUAL FRAMEWORK

Since our study inquired into the values of STEM scholarship grant programs to students' personal and professional being based on their storied experiences, the conceptualizations of value, story, experience, and identity form the four theoretical pillars underpinning our discussion.

Value

As readers may already have intuited, values can be in objects but are not "object-bound" (Rokeach, 1973, p. 17). Schwartz (1996) conceptualized "values as desirable, transmissional goals, varying in importance, that serve as guiding principles in people's lives" (p. 2). Different types of values (Schwartz, 1992; Schwartz et al., 2012) form a motivational continuum which helps to explain individuals' decision-making, attitudes, and behaviors. Those values do not exist separately. Instead, they may relate to each other in consistent or confrontational ways (Schwartz et al., 2012). For instance, seeking one's personal success may hinder one's actions to improve the welfare of others. Inspired by Schawtz's theories of value, we, instead of merely focusing on the economic value of scholarship programs, consider value as "multifaceted" (Rokeach, 1973, p. 2) in that it relates to students' past, current, and future; students' belief, thinking, and conduct; and students' decision-making concerning what is important (Schwartz, 1992; Schwartz & Bilsky, 1987, 1990).

Meanwhile, value "in one way or another" relates to "meaning" (Adler, 1956, p. 274). Through "observation" and "introspection" (Adler, 1956, p. 274), people discover value "in phenomena" and "human behaviors" (Adler, 1956, p. 273). Put differently, values are not just "objective, social elements" imposed on "individuals as given" (Thomas & Znaniecki, 1958, p. 1831), but they interact generatively with individuals' meaning-making of their lived experiences. This perspective of value enabled us to avoid viewing STEM scholarship programs from a detached point (Greene, 1995). Rather, we take students' "hearts and minds" (Goodson, 2007, p. 146) into consideration in our narrative inquiry into the influences of scholarships on students.

Experience

Students enter teaching–learning situations with "different personal histories which confer on [them] widely varying wants and capacities for satisfaction" (Schwab, 1959/1978, p. 172). Therefore, we do not view the seven students in our study simply based on which gender, race, and social class category they fall, but as students with particular experiences. Based on Dewey's (1938) philosophy of experience, students' actions and inner feelings unfold along an experiential continuum. "Each point [on the continuum] has a past experiential base and leads to an experiential future," according to Clandinin and Connelly (2000, p. 2). Also, experience as a whole is a "changing stream…characterized by continuous interaction of human thought with [the] personal, social, and material environment" (Clandinin & Rosiek, 2007). Therefore, we consider the NSF scholarship grants as a point of the students' experiential continuum, which connects with students' past and influences their future lives. Meanwhile, what happened outside the program is taken as important as well as what is achieved inside the program (Schwab, 1970). To sum up, it is within those various admixtures that we trace the clues of the value(s) of the scholarship-bearing programs to students.

Story

According to Clandinin and Connelly (2000), "experience happens narratively" (p. 19). "Human [beings] are storytelling organisms who, individually and socially, lead storied lives" (Connelly & Clandinin, 1990, p. 2). When the students in our study talked about the value of the scholarships, they shared stories about their families, teachers, peers, and anticipated careers with us. Thus, those students' stories are "portal[s]" (Connelly & Clandinin, 2006, p. 375) through which students find words to communicate their experiences, to make sense of their relationships with others, and to construct the meaning of their experiences. In the students' living, telling, retelling, and reliving of their stories (Connelly & Clandinin, 1990, p. 4), the connections between scholarship-bearing programs and students' lives in varied social milieus are revealed, which in turn point students in the direction of their imagined future.

Identity

The theme of student identity rose to the fore in the students' shared stories of change, because their narratives chronicle how students become who they are as human beings. In Palmer's (1998) words, students' identities are

> …an evolving nexus where all the [streams of experience] that constitute life converge in the… self: genetic makeup, the nature of the man and woman who gave life, the culture in which [students are] raised, people who have sustained [students] and people who have done [them] harm, the good and ill [they] have done to others and to [themselves], the experience of love and suffering-and much, much more. In the midst of this complex field, identity is a moving intersection of the inner and outer forces that make [youths] who [they are] … (p. 13).

Through storytelling, our student participants reflectively constructed and reconstructed how their experiences with the scholarship programs shaped who they are in their families, their communities, the STEM profession, and even the society. The students, with the fresh meaningfulness gleaned from their having of the scholarship, started their new "stories to live by" (Clandinin & Connelly, 1998, p. 149), which is the narrative inquiry term for identity. Additionally, our study reveals – from both contextual and individual perspectives – how students, through claiming their self-authorship, decided what to value in situations of uncertainty and ambiguity (King, 2003). Those "moments of choice" (Schwab, 1983, p. 245) revealed how students' self-images "are in the making" (Greene, 1995) personally, professionally, and socially.

RESEARCH METHOD

Narrative Inquiry

We employ narrative inquiry in our study to construct and reconstruct students' experiences via stories (Connelly & Clandinin, 1990; Dewey, 1938). The metaphorical "three-dimensional narrative space" (p. 49) conceptualized by Clandinin and Connelly (2000) acted as our framework. This space consists of temporality (continuity), personal/social (interaction), and place (situation/context). In other words, we explore students' experiences "over time, in a place or series of places, and in social interaction with milieus" (Clandinin & Connelly, 2000, p. 20) within and outside the NSF scholarship-bearing programs. Students' inner feelings are also attended to. In this way, students' experiences are unpacked while taking both "integrity and particularity" (Greene, 1995, p. 10) into consideration.

Data Sources

We as a research team have been doing studies on the experiences of students in the six aforementioned NSF grant programs. Data are collected from regular individual and focus group interviews with all the enrolled students. Based on the carefully transcribed data, we retrieved four major themes. One of the themes is the value of scholarship grants to students intending to study the STEM disciplines and to pursue STEM careers, which is the chosen topic of this study. This is also true of students of all races who are the first generation in their families to attend colleges. Furthermore, according teachers (Craig, Evans, Verma, Stokes, & Li, 2018/reprinted in this book), professors (Li et al., in press), and parents (Craig, Verma, Stokes, Evans, & Abrol, 2018/reprinted in this book) on the students in the grant programs.

Interpretive Tools

Three analytical tools – broadening, burrowing, and storying and restorying (Connelly & Clandinin, 1990) – helped our research team to seam together the narrative evidence we gathered. Using broadening, we situated the participants' narratives against the backdrop of national policies regarding STEM profession, the education as a whole, and the instrumentalism orientation, as discussed in the introduction and literature review sections of this chapter. Burrowing, the second research tool, enabled us to locate ourselves amid students' stories of their past, present, and imagined futures. Within the "continuity" and "wholeness" of an individual student's experiences (Clandinin & Connelly, 2000, p. 17), we were able to access to the "situation-specific understandings, the unmeasurable, and the unique" (Greene, 1995, p. 11) regarding the values of scholarships to students. The third analytical tool, storying and restorying, allowed us to capture changes in students' sense making of what to know, how to know, and what to believe in reconstructing their identities. Those up-close, human meanings (Goodson, 2007) led us to see how the course of students' lives can be potentially changed in their interactions with the scholarship-bearing programs. Taken together, the three interpretive devices helped us to deeply examine how our research participants' experiences with the programs interwoven with their personal and professional identities in their unfurling STEM academic and career trajectories. Data interpretations were member-checked (Lincoln & Guba, 1985) with participant students in follow-up interviews or email communications to ensure the trustworthiness of this narrative study.

Research Participants

In our ongoing research on the six NSF programs, each student talked about the value of scholarships to them in their interviews. In this chapter, we introduce readers to seven of the students: Omid Kassen, Anton Ivanov, Joyce Harding, Duong Pham, Sunjay Ritzvi, Kadeem Bello, and Tonya Goodson. Those students' experiences were chosen to illuminate multiperspectival insights into the influences of scholarships on their lives, careers, and selves. These pseudonyms were chosen to protect the identities of the students. Omid and Sanjay were undergraduate students who received S-STEM scholarships through the NSF computer science grant, while Anton, as a graduate student, received the SFS scholarship with a cybersecurity focus. Joyce and Duong were supported by the NSF Noyce Grants. As for Tonya, she was supported by two scholarship programs, one through the awarded S-STEM grant in the department of physics and another through a more recent Noyce grant award in *teach*HOUSTON program. Other than American-born Joyce, the students are first-generation Americans whose parents emigrated from countries like Iran, Russia, and Vietnam and from the continent of Africa. Each student has a different life journey before entering their STEM programs, which contributed to the different assessments of value they derived from their scholarship experiences, as showcased by the students' small stories (Georgakopoulou, 2004; Olson & Craig, 2009).

We feature the seven students' experiences in the form of narrative exemplars. According to Lyons and LaBoskey (2002), narrative exemplars are "concrete examples… elaborated so that members of a relevant research community can judge for themselves their 'trustworthiness' and the validity of observation, interpretations, etc." (p. 20). Put differently, we leave it to readers to decide whether our interpretation of the value of scholarship grants in this study adds to the knowledge base and if the findings we feature apply to their milieus and to possible actions they might take as educators in the future.

SMALL STORIES OF SCHOLARSHIP GRANT EXPERIENCES

The seven students' narratives of experience will now be presented. The students' small stories – told in their own terms and charged with their own emotions – revolve around (1) a blueprint of a STEM life, (2) a profession after one's first job disappeared, (3) a pathway to one's "best-loved self," (4) a life in a family-like community, (5) a reason for hope, (6) a bridge to overcome chasms, and (7) a call to teach.

A Blueprint of a STEM Life – Omid Kassen

Omid Kassen was a junior who was supported by the S-STEM computer science program. Omid immigrated to the United States from Iran with his family when he was a teenager. In a Persian neighborhood located in the fourth largest city in America, Omid's father works as a cashier and his mother as a hairstylist. At the time of their arrival, none of his family members "knew any English." Thankfully, Omid had high school teachers who supported him during his challenging transition from one country and culture to another. For instance, his ESL teacher created opportunities for Omid to speak in Farsi (Omid's home language) while learning English. With his increasing proficiency in English, Omid was then able to transfer his strength in Mathematics and Science (scoring in the top 10%–15% in his home country) to his academic learning in his high school in the United States. Also, his science teacher informed Omid of the S-STEM scholarship opportunity at the local research-intensive university, and Omid applied for the award.

The principal investigator of the grant then "took [Omid] under his wing." The scholarship paid for Omid's college tuition, which was an expense his immigrant family could not afford. As a first-in-family-to-college student, Omid was exposed to "eye-opening" experiences through research and professional conferences. Omid's interactions with others in the computer science community helped him to see from the inside "what computer science is" and "what it might be in [his] future." Furthermore, in the throes of his scholarship program experiences, Omid developed an avid interest in cybersecurity and software development. This may pave the way for him to pursue a graduate degree in Computer Science sponsored by the SFS program described earlier, which is held by the same STEM principal investigator.

A Profession When the First One "Disappeared" – Anton Ivanov

Anton Ivanov moved from Russia to Houston with his family members when he was three years of age. Currently, Anton is in his third year of study in the SFS program. He already has a BS in Chemical Engineering and a Master's degree in Geophysics. Anton previously worked for an oil company as a geophysicist. His master's degree education in Geophysics was sponsored by his employer. However, when the oil industry crashed in 2015, Anton lost his job. His younger sister was then taking the Medical College Admission Test (MCAT), and Anton also decided to apply. He received acceptable test scores to enter the medicine field but he did not meet the GPA admission requirement. Later, Anton saw the recruitment announcement for the SFS scholarship program. To Anton, his previous job involved writing "simple computer programs," but he had always wanted to write more "complicated" ones, which was a goal this cybersecurity program could help Anton to achieve. Therefore, Anton grabbed this opportunity presented to him. The program enabled Anton to retool himself in a related but different STEM profession. Sponsored trips to cybersecurity conferences additionally have helped Anton to envision his future career of "working for NASA and hacking terrorists."

A Pathway to "Best-Loved Self" – Joyce Harding

Joyce Harding, a Noyce scholarship recipient, is in the student teaching phase of the *teach*HOUSTON program. Joyce's decision to become a teacher was influenced by the caring she received from "great teachers" during her teenage years when she encountered "dark times" with her family. Joyce wants to return the caring she received to other children who are living in poverty like she did, which aligns beautifully with the Noyce goal of preparing STEM teachers for high-needs urban schools.

Growing up in a family adhering to fundamentalist religious beliefs, Joyce remembered her Dad saying such things as "Your Mother does not know anything because she comes from Eve." Gradually, Joyce found the knowledge she acquired at school "contradict[ed]" what she learned from her father's seemingly subtractive view of her mother at home. Moreover, with her teachers' encouragement to probe what she wanted to know, Joyce realized she could explore the unknown and make judgments based on her own critical thinking. Later, when she entered the Noyce scholarship program, Joyce became exposed to science taught as inquiry. This approach refreshed Joyce's perspectives of science as a learner herself and developed her strategies of inspiring students to critically think through constructive questioning and hands-on experiences. Meanwhile, as a female in the field of science, Joyce resisted the taken-for-granted belief that "males are better than females in science," a position directly articulated by one of her male professors. Not only did Joyce, role-modeling her African American female professor, actively speak her voice in STEM classes, but also she was committed to fostering girls' confidence and agency in STEM learning in the future as a teacher.

A Life in a Family-like Community – Duong Pham

Duong Pham, a mathematics major, is also a Noyce scholarship recipient in the *teach*HOUSTON program. *teach*HOUSTON has a balanced ratio of students representing different ethnicities. Despite their human differences, the students "form into one" – to borrow Duong's words and image. In Duong's view, *teach*HOUSTON "is like a family." This is the "flipside" of what he experienced in Vietnam and in other learning programs in which Duong had been enrolled in the United States.

Duong further explained: "When I go home, my mom take[s] care of my food. She give[s] me energy to go to school. When I go to school, my master teachers take care of me." Those teachers "held [Duong's] hands" and "helped [Duong]" along the way. For instance, when a Stanford Research Lab internship position became available, Duong did not think that he was ready for this opportunity. With the encouragement and support of his professors, Duong applied for it and was chosen for the competitive position. Duong was then able to work with Stanford University scientists for three months and to produce a publication based on the project he worked on, a skill that complemented his ability to contribute to the writing of NSF grant proposals for *teach*HOUSTON. Duong concluded:

> I learn[ed] a lot from them [*teach*HOUSTON faculty] from a science perspective and also as a person. They taught me to believe in myself. And I learned that I am going to fail a lot, and I am going to learn a lot from failures.

When asked about his future intentions, Duong shared that he planned to advance his education at another research-intensive university in New York City. In Duong's case, the scholarship experiences gave him a story to live by (Clandinin & Connelly, 1998) when Duong did not have a story of his own and did not believe that he could have one because of his immigrant background.

A Reason for Hope – Kadeem Bello

Kadeem Bello is an undergraduate student enrolled in the S-STEM computer science scholarship program. In many interviews, Kadeem shared intimate details of his "complicated life." Growing up in Nigeria, Kadeem was bullied at school because he was thin and wore glasses, which were viewed as signs of weakness in his local culture. Also, Kadeem's parents believed in "spare the rod, spoil the child." Bullying at school and strict punishment at home led to Kadeem becoming depressed.

When Kadeem reached the age of 18, he started "questioning everything" that he had known. Then he decided to travel to the United States to seek a new way of living. In Chicago, Kadeem worked and studied hard to become independent of his parents. However, Kadeem overworked himself and became clinically diagnosed with depression. While he was struggling with his mental health issues, a white female professor who Kadeem met on his flight from Africa brought him into her family. Kadeem then left Chicago for Houston to continue his course of study. However, not long after he settled in the southern state, his car became damaged beyond repair in a 100-year flood. In order to save money, he moved out of his apartment and lived in the basement storage space of his friend's apartment. Kadeem recalled his sad state of affairs during that period of time:

> I was going through much soul searching in those days. So, I questioned so many things, and I felt like I needed to change. I felt like the only reasonable way for me to really live in this world was if I did things more realistically, more logically. You know, if I was more rational about things rather than just some belief system that, you know, my parents told me… Like, I never really had any real reason of my own why I'm doing this or that, you know.

Fortuitously, Kadeem became introduced to the S-STEM scholarship program. For him, computer science had the logical features that he had been seeking. He applied for the scholarship and received it in the middle of a semester. "It [the scholarship] gave me hope," he said

> Over the past year, I was pretty much sleeping in [a] storage unit…that is, until just last week. The scholarship allowed me to be able to hang in there; I found hope in knowing that something is coming.

In follow-up interviews, Kadeem went on to say that "the scholarship heightened [his] sense of compassion." He talked about how he planned to bring learning resources to African children in need. The compassion and hope afforded Kadeem Bello via the scholarship program, in turn, seeded and expanded Kadeem's compassion for others. No longer depressed, he sees himself as a "wounded healer and ambassador," one who can be an agent of change who can bring hope to others.

A Bridge to Overcome Chasms – Sunjay Ritzvi

Sunjay Ritzvi grew up in a Pakistani family that moved to America in his early childhood years. In Sunjay's words, becoming an engineer was "kind of engraved in [his] mind since the day [he] [was] born." Put differently, Sunjay entered the

STEM field living a story that had been given to him since birth by his family members. The S-STEM scholarship enabled Sunjay to "focus more on [his] education." For instance, Sunjay took more than 20 credit hours so that he could keep his graduation plan on track. When asked about his program experiences, Sunjay replied:

> I think it helps that it [the program] gives you more freedom in terms of, like, what you want to study or, like, what you want to learn. So, in that sense you're more able to gear yourself towards, like, your career path...It feels like it made it easier for me to make a decision on the career path I'm on, and just, like, able to freely study towards it.

The conversation continued this way:

> Interviewer: It kind of cleared away any uncertainty?
>
> Sunjay Ritzvi: Like I said, it's like being at the end of the college. Like, at the beginning of college, you're just taking introduction classes, and it seems very unclear why – like, a lot of times when I was a freshman, I would say stuff to myself, like, how does this help or, like, you know – like, when people are in, like, in math course they'll say something, like, when are you going to use this in real life? But, I mean, the farther you go, the more you realize how much it [awarded scholarship] has helped.

For Sunjay Ritzvi, his receipt of the scholarship bridged theory–practice chasms that he had experienced earlier in his education. Sunjay now felt certain he can confidently begin his Computer Science career without having to deal with the induction shock that beginners often encounter in their first place of employment.

A Call to Teach – Tonya Goodson

Tonya Goodson's parents identify as West African Americans, but Tonya simply considers herself African American. In Tonya's words, "I am the first person in my family and extended family to go all the way through the American school system. That is why I comfortably say I am African American." Tonya immigrated to Texas several years ago with her mother, father, and two younger brothers, one who is disabled. Academically, she has been a fine student, one who believes that learning – not grades – is what education is all about. As Tonya maintained, she "learns to learn, not learns for grades...." "A hard-earned C" is always better than "a frivolous A," she opined.

Tonya's receipt of two scholarship grant awards immediately made life "less stressful" for her family and her. Tonya shared:

> My mother, who was a teacher in my home country, retrained herself as a pharmacist when we came to the US. However, my family regularly sends money to our less fortunate relatives in Africa. Also, my disabled brother's medications are very expensive...and they can now be more easily purchased. So, receiving the scholarships was awesome not only for the financial benefits, but also because they relieved the stress of both my family members and me. I am extremely happy.

Prior to attending the Noyce Scholarship Conference in Alabama, Tonya Goodson was on a premedicine career path and admittingly had little interest in being a teacher despite taking *teach*HOUSTON classes as her capstone. As part of her studies, she enrolled in classes focusing on culturally relevant pedagogy and the need for social justice. Those classes peaked her interest and caused her to "fall in love with teaching." However, what made her fully commit to the teaching profession happened at the Noyce leadership conference she attended in Alabama. While listening to many strong African American keynote speakers and conference organizers and sharing her own poster about using technology to increase student engagement, she experienced a complete paradigm shift:

> I recognized that I was the student the speakers were talking about. I was the student the teachers did not believe in…The student they were speaking about was me… I then recognized that I could create an atmosphere that would be better for students and a place where they could grow…I could help the homeless children coming into the schools. I could be their advocate.

This full-blown awakening led Tonya to realize that teaching – not medicine – was her calling (Bullough, Jr. & Hall-Kenyon, 2012; Dewey, 1929). Tonya knew that teaching would not just be something she would do to repay a grant commitment, but a calling to which she would be committed for many, many years to follow.

DISCUSSION

Distilled to the essence, the seven small stories of experience that the undergraduate and graduate STEM students lived and told – and relived and retold – revolved around three overarching themes regarding the value of the scholarships: (1) filling a void in their academic, personal, and/or professional lives, (2) enabling students to discover new selves, and (3) human preparation.

Filling a Void in Students' Academic, Personal, and/or Professional Lives

All of the stories we recounted identified a gap or a void that needed to be filled in the students' academic, personal, or professional lives. The NSF scholarships addressed those gaps in varied ways. The S-STEM scholarship provided Omid with access to higher education. This enabled Omid to continue developing his strength in STEM and to envision his future in the STEM field with a broader view in mind. Anton, because of the SFS scholarship, found a new profession when he could not foresee a future in either the oil industry or the medical profession. It also helped Anton to avoid placing a huge financial burden on his family members or himself. For Joyce, the Noyce scholarship offered her a new inquiry approach to learning, teaching, and life. In the *teach*HOUSTON program, Duong found a diverse community where he felt he belonged as a former immigrant to the country. This community mentored him to become a qualified STEM teacher and researcher. The S-STEM scholarship gave Kadeem Bello "a new story to live by." This freed him from the stuck story in which he felt mentally and physically trapped. For Sunjay Ritzvi, who could have collapsed from the crushing parental pressure forcing him to become an engineer, the

STEM scholarship provided him with a smooth trajectory into the cybertechnology profession, which his family members subsequently approved of.

Enabling Students to Discover Their New Selves

Each of the students' experiences with the scholarship grants displayed a journey of self-discovery, that is, a reconstruction of their identities. For instance, Omid began to imagine himself as a STEM researcher focusing on his newly developed interests in the field. Joyce transformed negative experiential fodder that had come her way into "educative experiences" (Dewey, 1938, p. 29) and geared herself to becoming a female scientist advocating for gender justice. The NSF scholarship program provided answers Kadeem Bello sought in his soul searching – something "more realistic, more logical." This helped Kadeem to resolve tensions deeply embedded in himself and to start making something of his own (Sennett, 1999, p. 138) in his learning, his life, and his future profession. Thanks to the advocacy of his *teach*HOUSTON family members, Duong's personal self grew in self-confidence and courage, which, in turn, encouraged Duong to explore more stories he desired to live but dared not enact in the academic and profession. Finally, for Tonya Goodson, the scholarship program classes and the minority role models she met at the Noyce conference in Alabama ignited a passion within her and enabled her to recognize that teaching had supplanted medicine as her calling in life. Those connections students discovered between the scholarship grants and their selves helped them to strengthen their own voices, to enrich their identities, and to understand themselves, people around them, and the world in ways different than before. This helped students to avoid "be[ing] blind to a larger picture" (Seymour, 2016, p. 6), a key idea that brings us to our third all-encompassing theme: human preparation.

Human Preparation

From the small stories of experience, we see how STEM education subtly blended with human preparation time and time again. For instance, Kadeem, with a "sense of compassion" heightening his STEM scholarly activities, decided to become a "wounded healer" to help other children in Africa, who might be facing similar life and mental challenges that he did as a child growing up in Nigeria. His narrative showed him "surrender[ing] to a story is to keep [his] very humanity alive" (Fagin as cited in Popova, 2018). For Joyce, while investing efforts in developing herself as a female equal to men in the STEM fields, she also equipped herself with knowledge and strategies to foster girls' STEM learning in her future teaching career. Duong Pham became a person who could resiliently pave his own way, while concurrently recognizing that failure also presents opportunities for learning and growth in life. Tonya strengthened her self-and-racial identity awareness and wanted to share the same pride in the STEM teaching profession as those of her recently claimed role models at the Noyce conference. Therefore, the scholarship grant programs not only intuitively taught students to be more understanding of themselves in the making but also cultivated students to become beings flourished with "an integral and human way of thinking" (Seymour, 2016,

p. 7). This lays a foundation for students to connect with themselves, others, and the world (Seymour, 2016), which give the students' lives new meanings.

CLOSING STATEMENTS

Efforts have been invested to build STEM programs through the integrated approach (Tsui, 2007) which attends to factors such as institutional leadership, faculty engagement, learning environment, and so on. Those top-down measures, reflecting good intentions, attempt to maximize the chances of meeting the needs of diverse students. However, our study revealed that the values of scholarships to students are not simply a "natural phenomenon which under careful study yield law like relationships among variables" (Tom, 1980, p. 317). Rather, they are rooted in unrepresented students' storied lives lived in relationships with people, places, and things in varied milieus. The scholarship programs necessarily become "inherently storied places" (Craig, 2020). Such matters easily go unnoticed when a single focus is placed on the grades and graduation rates in STEM programs.

Also, we see how the underrepresented students, influenced by the dynamics within their individual experiences and the programs, not only survived but also thrived in the programs, in their lives, and very possibly in their future STEM career. Each of the students came to the programs with their own challenges, as illustrated in their individual experiences. The scholarship-bearing programs helped the students to resolve their challenges through varied ways such as providing students opportunities for them to build on their strengths and cultivating a resilient mindset within students. Meanwhile, beyond considering themselves as learners who just fulfilled the requirements of the program, the students meaningfully engaged themselves with the STEM content, peers, faculties, program activities, communities, and their own culture. In the journeys of discovering and refining who they are, what they want to become, and how to relate to others, the students found their own ways for being and becoming better persons.

Furthermore, in our narrative inquiry research, we discovered how the scholarship programs organically intertwined STEM education with humanity. We saw the students advocating for humanity in their interactions with others, even when a foundation for developing intellectual, civic, and practical skills and dispositions was not necessarily part of the communicated "ends" of their STEM-related curriculum. Their discovery of "deeper selves" (Seymour, 2016, p. 5) provided students with routes to congeal new meanings of experiences, which further connected them with large social issues (community, gender, race) needing attention. These experiences helped students to seek "a more sustainable and meaningful way of being and thinking" (Seymour, 2016, p. 2) and cultivated in them "an attitude of a mind liberated from apathy and self-will, an attitude of mind [that combined] life, work, duty and the realities of belief" (Sadler, 1930, p. 256). These outcomes were an added boon that complemented – not detracted from – NSF's original goal of increasing qualified STEM workers for the nation.

In summary, current approaches to education, especially STEM education, have been critiqued for focusing on "skills and knowledge [students] needed to work and function at an 'economically sound' level in society" (Seymour, 2016, p. 4). Consequentially, the "economic being, as opposed to the human being," is prioritized (Seymour, 2016, p. 4). Our inquiry into the underrepresented students' experiences unearths how STEM education can relate to multiple significant dimensions of life (Seymour, 2016). Who the students are is just as important as "what [they] can know and do" (Seymour, 2016, p. 8). Those students established their sense of value in autonomous and committed ways while promoting their personal welfare and seeking the common good of others (Thornburg, 2008). Thus, we suggest scholarships be viewed as financial incentives that are "overlaid with purpose, passion, and meaning" (Goodson, 2007, p. 139) rather than instrumentalized material rewards aimed at technical delivery. With those enriched perspectives in this study, our long-term hope is to navigate readers into new spaces of inquiry with regard to preparing and sustaining students of diversity in STEM education and also in the broad context of higher education, personally, professionally, and socially.

AUTHORS' NOTE

This research was funded by the following NSF grants:

- Noyce Grant: *Recruitment, Preparation and Retention of STEM Students as High School Teachers* [1240083]. Principal Investigator: Dr. Donna W. Stokes; Dr. Paige K. Evans
- S-STEM Scholar Program: *Preparing Computer Science Students for Global Challenges of the 21st Century* [1356705]; Principal Investigator: Dr. Rakesh Verma
- CyberCorps (R): *Scholarship For Service (SFS): Increasing Talented Trusted Professionals* [1433817]. Principal Investigator: Dr. Rakesh Verma
- Noyce Grant Track 1: *STEM Teacher Preparation at the University of Houston: Learning through Informal and Formal Experiences* (UH-LIFE) [1557309]. Principal Investigator: Dr. Paige K. Evans
- STEM Scholarship Program: *Promotion and Retention of STEM Education through Networking Team (PARENT) Support* [1644191. Principal Investigator: Dr. Donna W. Stokes
- Noyce Grant Track 3: *Leadership through Advocacy and Development in STEM* (UH-LEAD) [1759454]. Principal Investigator: Dr. Paige K. Evans

REFERENCES

Adler, A. (1956). *The individual psychology of Alfred Adler*. New York, NY: Basic Books.

Arendt, H. (1998). *The human condition*. Chicago, IL: University of Chicago Press.

Beane, J. A. (1998). Reclaiming a democratic purpose for education. *Educational Leadership*, *56*(2), 8–11.

Bullough, R., Jr, & Hall-Kenyon, K. (2012). On teacher hope, sense of calling, and commitment to teaching. *Teacher Education Quarterly*, *39*(2), 7–27.

Buykx, P., Humphreys, J., Wakerman, J., & Pashen, D. (2010). Systematic review of effective retention incentives for health workers in rural and remote areas: Towards evidence-based policy. *Australian Journal of Rural Health*, *18*(3), 102–109.

Carpenter, D. M. (2005). Presidential rhetoric and the purpose of American education. *The Educational Forum*, *69*(3), 278–290.

Chang, M. J., Eagan, M. K., Lin, M. H., & Hurtado, S. (2011). Considering the impact of racial stigmas and science identity: Persistence among biomedical and behavioral science aspirants. *The Journal of Higher Education*, *82*(5), 564–596.

Christensen, R., Knezek, G., & Tyler-Wood, T. (2014). Student perceptions of science, technology, engineering and mathematics (STEM) content and careers. *Computers in Human Behavior*, *34*, 173–186.

Clandinin, D. J., & Connelly, F. M. (1998). Stories to live by: Narrative understandings of school reform. *Curriculum Inquiry*, *28*(2), 149–164.

Clandinin, D. J., & Connelly, F. M. (2000). *Narrative inquiry: Experience and story in qualitative research*. San Francisco, CA: Jossey-Bass.

Clandinin, D. J., & Rosiek, J. (2007). Mapping a landscape of narrative inquiry: Borderland spaces and tensions. In D. J. Clandinin (Ed.), *Handbook of narrative inquiry: Mapping a methodology* (pp. 35–75). Thousand Oaks, CA: Sage Publications.

Clewell, B. C., & Ficklen, M. S. (1987). Effective institutional practices for improving minority retention in higher education. *Journal of College Admission*, *116*, 7–13.

Connelly, F. M., & Clandinin, D. J. (1990). Stories of experience and narrative inquiry. *Educational Researcher*, *19*(5), 2–14.

Connelly, F. M., & Clandinin, D. J. (2006). Narrative inquiry. In J. Green, G. Camilli, & P. Elmore (Eds.), *Handbook of complementary methods in education research* (pp. 375–385). Mahwah, NJ: Lawrence Erlbaum.

Craig, C. (2020). *Curriculum making, reciprocal learning and the best-loved self.* New York, NY: Palgrave Pivot.

Craig, C., Evans, P., Verma, R., Stokes, D., & Li, J. (2018). A tribute to 'unsung teachers': Teachers' influences on students enrolling in STEM programs with the intent of entering STEM careers. *European Journal of Teacher Education*, *42*(3), 1–24.

Craig, C., Verma, R., Stokes, D., Evans, P., & Abrol, B. (2018). The influence of parents on undergraduate and graduate students' entering the STEM disciplines and STEM careers. *International Journal of Science Education*, *40*(6), 621–643.

Dewey, J. (1916). *Democracy and education*. New York, NY: Macmillan.

Dewey, J. (1929). *My pedagogic creed.* New York, NY: Progressive Education Association.

Dewey, J. (1938). *Experience and education*. New York, NY: Basic Books.

Diekman, A. B., & Benson-Greenwald, T. M. (2018). Fixing STEM workforce and teacher shortages: How goal congruity can inform individuals and institutions. *Policy Insights from the Behavioral and Brain Sciences*, *5*(1), 11–18.

Dortch, D., & Patel, C. (2017). Black undergraduate women and their sense of belonging in STEM at predominantly White institutions. *NASPA Journal About Women in Higher Education*, *10*(2), 202–215.

Fenske, R. H., Porter, J. D., & DuBrock, C. P. (2000). Tracking financial aid and persistence of women, minority, and needy students in science, engineering, and mathematics. *Research in Higher Education*, *41*(1), 67–94.

Georgakopoulou, A. (2004, May). *Narrative analysis workshop: How to work with narrative data.* Paper presented at Narrative Matters 2004: An Interdisciplinary Conference on Narrative Perspectives, Approaches, and Issues across the Humanities and Social Sciences, Fredericton, New Brunswick.

Giroux, H. A. (1988). *Schooling and the struggle for public life: Critical pedagogy in the modern age.* Minneapolis, MN: University of Minnesota Press.

Goodson, I. (2007). All the lonely people: The struggle for private meaning and public purpose in education. *Critical Studies in Education*, *48*(1), 131–148.

Graham, Y. (2017). *Examining the effect of a scholarship and living-learning program intervention on first-generation college student thriving*. Doctoral dissertation. Baylor University, Waco, Texas.

Greene, M. (1995). *Releasing the imagination: Essays on education, the arts, and social change*. San Francisco, CA: Jossey-Bass.

Harris, C. M. (2015). *Surviving the sciences: Factors that influence exit from the STEM workforce*. Baltimore County, MD: University of Maryland.

Hartoonian, H. M. (1999). Education for sale: What's a democratic principle worth? *Social Education, 63*(4), 241–243.

Hungerford, C., & Hodgson, D. (2013). Addressing mental health nursing workforce issues in Australia: A case study analysis. *The Journal of Mental Health Training, Education and Practice*, *8*(2), 89–102.

King, T. (2003). *The truth about stories*. Toronto, CA: Anansi Press.

Lawrenz, P., Appleton, J., Bullitt Bequette, M., Desjardins, C., Liou, P.-Y., Madsen, C. (2008). University of Minnesota Evaluation of the Robert Noyce Teacher Scholarship Program, Final report section one: Planning and survey data, Noyce Program Evaluation. University of Minnesota, Minneapolis.

Li, J., Evans, P. K., Craig, C. J., Stokes, D. W., Verma, R., & Zhu, G. (in press). The influence of professors on students enrolled in the STEM programs with the intent of embarking on STEM careers. In C. J. Craig., P. K. Evans & D. W. Stokes (Eds.), *Preparing teachers to teach the STEM disciplines in America's urban schools*. Bingley, UK: Emerald Publishing Limited.

Lincoln, Y., & Guba, E. (1985). *Naturalistic inquiry*. Beverly Hills, CA: Sage.

Liou, P. Y., Kirchhoff, A., & Lawrenz, F. (2010). Perceived effects of scholarships on STEM majors' commitment to teaching in high need schools. *Journal of Science Teacher Education*, *21*(4), 451–470.

Liou, P., & Lawrenz, F. (2010). Optimizing teacher preparation loan forgiveness programs: Variables related to perceived influence. *Science Education*, *95*(1), 121–144.

Lyons, N., & LaBoskey, V. K. (Eds.). (2002). *Narrative inquiry in practice: Advancing the knowledge of teaching*. New York, NY: Teachers College Press.

Malcom, S. M., & Malcom-Piqueux, L. E. (2013). Critical mass revisited: Learning lessons from research on diversity in STEM fields. *Educational Researcher*, *42*(3), 176–178.

Marginson, S., Tytler, R., Freeman, B., & Roberts, K. (2013). *STEM country comparisons: International comparisons of science, technology, engineering and mathematics (STEM) education*. Melbourne: Australian Council of Learned Academies.

Masterton, J., Moss, D., Korin, S., & Watters, D. (2014). Evaluation of the medium-term outcomes and impact of the Rowan Nicks scholarship programme. *ANZ Journal of Surgery*, *84*, 110–116.

Miller, R. A., Vaccaro, A., Kimball, E. W., & Forester, R. (2020). "It's dude culture": Students with minoritized identities of sexuality and/or gender navigating STEM majors. *Journal of Diversity in Higher Education*. doi:10.1037/dhe0000171.

Museus, S. D., Palmer, R. T., Davis, R. J., & Maramba, D. C. (2011). *Racial and ethnic minority students' success in STEM education*. Hoboken, NJ: Jossey-Bass.

National Science Foundation, National Center for Science and Engineering Statistics. (2013). *Women, minorities, and persons with disabilities in science and engineering: 2013*. NSF 13–304. Arlington, VA. Retrieved from http://www.nsf.gov/statistics/wmpd/.

Olson, M., & Craig, C. (2009). Small stories and meganarratives: Accountability in balance. *Teachers College Record*, *111*(2), 547–572.

Ost, B. (2010). The role of peers and grades in determining major persistence in the sciences. *Economics of Education Review*, *29*(6), 923–934.

Palmer, P. (1998). *The courage to teach: Exploring the inner landscape of a teacher's life*. San Francisco, CA: Jossey-Bass.

Phelan, A. M. (2009). A new thing in an old world? Instrumentalism, teacher education and responsibility. In F. J. Benson, & C. Riches (Eds.), *Engaging in conversation about ideas in teacher education* (pp. 105–114). New York, NY: Peter Lang.

Pinar, W. F. (2012). *What is curriculum theory?* New York, NY: Routledge.

Popova, M. (2018, December 18). A 100-year-old Holocaust survivor on how books save lives. Retrieved from https://www.brainpickings.org/2018/12/18/a-velocity-of-being-helen-fagin/.

Riegle-Crumb, C., King, B., Grodsky, E., & Muller, C. (2012). The more things change, the more they stay the same? Prior achievement fails to explain gender inequality in entry into STEM college majors over time. *American Educational Research Journal*, *49*(6), 1048–1073.

Rodriguez, S. L., & Blaney, J. M. (2020). "We're the unicorns in STEM": Understanding how academic and social experiences influence sense of belonging for Latina undergraduate students. *Journal of Diversity in Higher Education*. doi:10.1037/dhe0000176.

Rokeach, M. (1973). *The nature of human values*. New York, NY: Free Press.

Sadler, M. (1930). The outlook in secondary education: A liberal education. *Teachers College Record*, *32*(3), 256.

Schreiner, L. A. (2010). The "thriving Quotient": A new vision for student success. *About Campus*, *15*(2), 2–10.

Schuurs, S., & Amsters, D. (2010). Evaluation of multiprofessional scholarships: The community rehabilitation workforce project experience. *Focus on Health Professional Education: A Multidisciplinary Journal*, *12*(1), 23.

Schwab, J. J. (1959/1978). The "impossible" role of the teacher in progressive education. *The School Review*, *67*, 139–159.

Schwab, J. J. (1970). *The practical: A language for curriculum*. Washington, DC: National Education Association.

Schwab, J. (1980). Social sciences. *The Center Magazine*, *13*(1), pp. 36–41.

Schwab, J. (1983). The practical 4: Something for curriculum professors to do. *Curriculum Inquiry*, *13*(3), 239–265.

Schwartz, S. H. (1992). Universals in the content and structure of values: Theoretical advances and empirical tests in 20 countries. In M. Zanna (Ed.). *Advances in experimental social psychology* (Vol. 25, pp. 1–65). Orlando, FL: Academic.

Schwartz, S. H. (1996). Value priorities and behavior: Applying a theory of integrated value systems. In C. Seligman, J. M. Olsen, & M. P. Zanna (Eds.), *Values: The Ontario symposium* (pp. 1–25). Hillsdale, NJ: Lawrence Erlbaum Associates.

Schwartz, S. H., & Bilsky, W. (1987). Toward a psychological structure of human values. *Journal of Personality and Social Psychology*, *53*, 550–562.

Schwartz, S. H., & Bilsky, W. (1990). Toward a theory of the universal content and structure of values: Extensions and cross-cultural replications. *Journal of Personality and Social Psychology*, *58*, 878–891.

Schwartz, S. H., Cieciuch, J., Vecchione, M., Davidov, E., Fischer, R., Beierlein, C., … Konty, M. (2012). Refining the theory of basic individual values. *Journal of Personality and Social Psychology*, *103*(4), 663–688.

Sennett, R. (1999). *The corrosion of character, the personal consequences of work in the new capitalism*. London: W. W. Norton.

Seymour, M. (2016). Educating for humanity: Rethinking the purposes of education. In M. Seymour, & H. M. Levin (Eds.), *Educating for humanity: Rethinking the purposes of education* (pp. 1–10). New York, NY: Routledge.

Seymour, E., & Hewitt, N. M. (1997). *Talking about leaving*. Oxford; Boulder, CO: Westview Press.

Smith, D. G. (2009). *Diversity's promise for higher education: Making it work*. Baltimore, MD: John's Hopkins University Press.

Smith, E. (2011). Women into science and engineering? Gendered participation in higher education STEM subjects. *British Educational Research Journal*, *37*(6), 993–1014.

Stout, J. G., Grunberg, V. A., & Ito, T. A. (2016). Gender roles and stereotypes about science careers help explain women and men's science pursuits. *Sex Roles*, *75*(9–10), 490–499.

Thomas, W. I., & Znaniecki, F. (1958). Introduction in Part IV: Life record of an immigrant. In W. I. Thomas, & F. Znaniecki (Eds.), *The polish peasant in Europe and America*. New York, NY: Dover Publications.

Thornberg, R. (2008). The lack of professional knowledge in values education. *Teaching and Teacher Education*, *24*, 1791–1798.

Tom, A. R. (1980). Teaching as a moral craft: A metaphor for teaching and teacher education. *Curriculum Inquiry*, *10*(3), 317–323.

Tsui, L. (2007). Effective strategies to increase diversity in STEM fields: A review of the research literature. *The Journal of Negro Education*, *76*(4), 555–581.

Wang, M. T., Ye, F., & Degol, J. L. (2017). Who chooses STEM careers? Using a relative cognitive strength and interest model to predict careers in science, technology, engineering, and mathematics. *Journal of Youth and Adolescence*, *46*(8), 1805–1820.

Worthington, R. L., Hart, J., & Khairallah, T. (2010). Counselors as diversity change agents in higher education. In J. G. Ponterotto, J. M. Casas, L. A. Suzuki, & C. M. Alexander (Eds.), *Handbook of multicultural counseling* (3rd ed., pp. 563–576). Thousand Oaks, CA: Sage.

WHERE ARE THE *teach*HOUSTON PRESERVICE CANDIDATES NOW? ARE THEY STILL IN THE URBAN TEACHER FORCE?

Paige K. Evans, Mariam Manuel, Ha Nguyen, Donna W. Stokes, Cheryl J. Craig, Xiao Han and Jeffrey Morgan

ABSTRACT

*This chapter traces the career trajectories of the teach*HOUSTON *science, technology, engineering, and mathematics (STEM) teachers since the inception of the program. It asks whether the National Science Foundation (NSF) investment of millions of dollars in STEM education produced more STEM teachers of high quality for the diverse, urban area. The chapter is filled with descriptive statistics and stories. Two major findings are that the teach*HOUSTON *program produced double-digit physics teachers when the Greater Houston area had not had a freshly prepared physics teacher in over a decade. Additionally, teach*HOUSTON *graduates have distinguished themselves by being named recipients of several awards such as beginning teacher of the year awards, district teacher of the year awards, among other distinctions. teach*HOUSTON *alumni are also serving in a variety of leadership capacities for high-need public school districts. The chapter ends with a discussion of the program's strengths as well as the areas in which it continues to grow. Three new NSF grants allow for continued improvement and transition this work into two additional books proposed in this three-volume series.*

Preparing Teachers to Teach the STEM Disciplines in America's Urban Schools
Advances in Research on Teaching, Volume 35, 201–226

ISSN: 1479-3687/doi:10.1108/S1479-368720210000035012

Keywords: STEM teacher preparation; inquiry-based learning; teacher retention; workforce; urban STEM teachers; *teach*HOUSTON

Decades of research support the assertion that teachers are the most important factor impacting the quality of education (i.e., Organization for Economic Cooperation and Development, 2005; Schleicher, 2018). During the critical years of K-12 education, teachers play a significant role in shaping the lives and characters of their students. This impact extends beyond the acquisition of academic skills and involves fostering students' self-esteem, which ultimately reinforces their motivation and learning. Given the power teachers hold to transform the lives and learning of their students, it is imperative that substantial resources are invested in the development and preparation of quality teachers (Darling-Hammond, 2006). Hence, university-based teacher preparation programs have been charged with equipping aspiring teachers with the tools and resources needed to teach the twenty-first century student.

The *teach*HOUSTON program through innovation in recruitment, retention and training has produced a highly qualified science, technology, engineering, and mathematics (STEM) teacher workforce. As readers will recall, *teach*HOUSTON started with 14 students in 2007 and now serves approximately 370 students. The goal was to increase the production of STEM teachers through an innovative model that included recruiting STEM majors to become certified teachers through obtaining teacher certification (Craig, Evans, Stokes, & Bott, 2017). Paramount to the success of the *teach*HOUSTON program are the essential pillars of success, including (1) strong collaboration between the Colleges of Natural Sciences and Mathematics (NSM) and Education; (2) active recruitment of NSM majors to take the two initial one-hour *teach*HOUSTON courses; (3) a focus on developing deep-level understanding of the subject material and incorporating effective approaches using technology in teaching; (4) guidance and inspiration provided by faculty and highly experienced public school teachers who serve as master teachers; (5) compact degree plans allowing students to complete their major and *teach*HOUSTON courses in four years; (6) courses taught by faculty members experienced in teaching in the public schools and actively engaged in STEM education research; (7) integrated research-based professional development courses that focus on teaching both mathematics and science; (8) discipline specific courses which include pedagogy; and (9) a wide array of informal student experiences and leadership opportunities through outreach efforts around the Greater Houston area (Craig et al., 2017).

In this chapter, the successes of the program will be highlighted, along with the statistics of the graduates including enrollment and graduation data as well as graduation and retention rates. A map of where the graduates are teaching including the impact to the local community will be presented. A description of the number of graduates entering teacher leadership positions/roles through the support of the program will also be discussed. Additionally, the stories of three *teach*HOUSTON STEM graduates will be featured. Lastly, potential areas for growth and the future plans to further build STEM teacher capacity of the program utilizing the new National Science Foundation (NSF) grants will be noted.

*teach*HOUSTON BY THE NUMBERS

In this section, a summary of *teach*HOUSTON by the numbers will be provided, along with program enrollment, major breakdown, STEM teacher production, gender and ethnicity, time to degree completion, and retention statistics. As previously indicated, prior to the 2007 launch of the *teach*HOUSTON program, the university's certification program produced an average of four STEM teachers annually. Subsequent to the receipt of the first NSF Grant, DUE # 1240083, the STEM teacher production increased, on average, to 32 per year during the 2013–2016 period. With the receipt of a second NSF Grant, DUE # 1557309, the production increased, on average, to 40 per year during 2016–2019. The program graduated 48 STEM teachers in 2019–2020 (Fig. 1). It is on target to produce approximately 70 teachers during the 2020-2021 academic year.

Program Enrollment

Over the past five years, the *teach*HOUSTON program enrollment has steadily increased, averaging 325 students per year. As per Fig. 2, the program enrolled 307 students in 2015–2016. In 2017–2018, the total active enrollment decreased to 278 students (or 9%). This is due to several things occurring in the 2012-2013 academic year including (1) lower program enrollment which ultimately resulted in a smaller graduation class in 2017–2018; (2) no longer reimbursing tuition for students taking the CUIN 1101 course; (3) the transition to operating the program without grant funding; and (4) the understaffing of faculty. However, with enhanced recruitment, human resources, and additional grant funding, the program's enrollment increased sharply in 2018–2019 from 278 to 368 students (or 32.4% increase). The active enrollment number was stable at 369 students for 2019–2020.

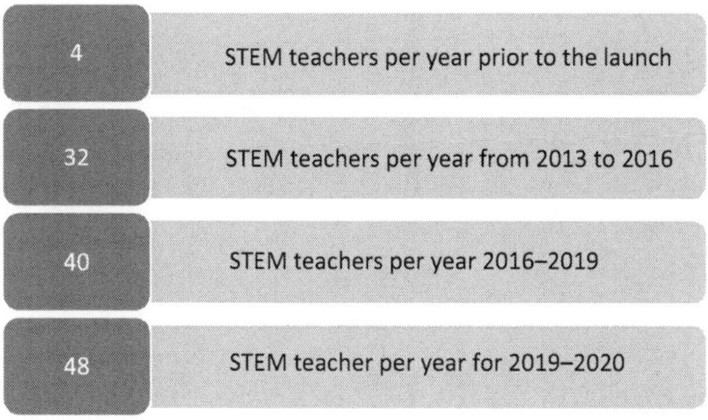

Fig. 1. *teach*HOUSTON STEM Teacher Production.

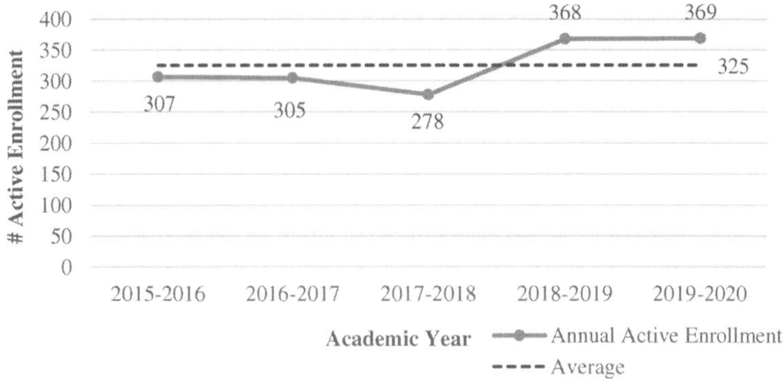

Fig. 2. *teach*HOUSTON Program's Annual Enrollment.

Enrollment by Major

The program certifies numerous STEM majors, which are identified in Fig. 3. Biology majors account for the largest percent of *teach*HOUSTON enrollment during 2015–2020, with, on average, 36%, followed by mathematics students (26%), while enrollment of chemistry, engineering, physics, computer science, and geology majors continues to be modest (combined 19%). This is in part due to the small pool of students for recruitment in those fields as compared to the number of biology and math majors in NSM. However, it is encouraging to observe the increase in computer science students enrolling in the *teach*-HOUSTON program, from only 5 students (or 1.6%) in 2015–2016 to 36 students (or 9.8%) in 2019–2020. This increase is due to a partnership with the computer science department through grant projects, enhanced efforts in recruitment, and the addition of a computer science education course in the *teach*HOUSTON program.

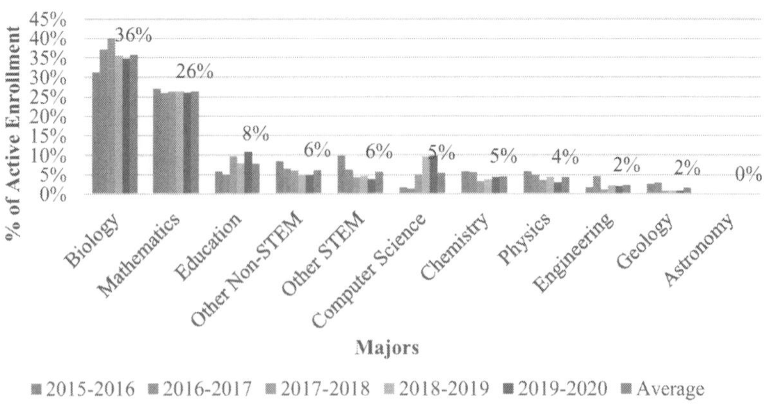

Fig. 3. *teach*HOUSTON Enrollment by Major.

Graduation Demographics

Although the overall production of US teachers fell 30% between 2009 and 2014 (U.S. Department of Education, 2016), the *teach*HOUSTON program experienced a strong increase in STEM teacher production. In fact, as of Spring 2020, the program has produced 347 graduates since its inception in 2007, therefore, making a critical contribution to the nation's draining pool of STEM teachers. Fig. 4 shows that the program's production of STEM teachers continues to rise. A key factor that contributed to the increase of teachers produced during the 2019-2020 school year was a new grant from the Department of Education, which resulted in the creation of a new certification track. The new track, the Alternative Certification Program for STEM (ACP for STEM at UH), targets students nearing or finishing their degrees and/or career changers and will be described later in this chapter.

Graduates by Major

Readers will recall that the *teach*HOUSTON program was formed to combat the staggering shortage of secondary STEM teachers across the United States and in Texas. Marder and Hamrock (2016) estimated that approximately 40% of mathematics teachers are teaching out of their fields and more than 60% of science teachers lack adequate qualifications in the United States. This plight is not much better in Texas where over 30% of middle school math and science teachers are teaching out of their fields, and 13.3% and 28.7% of high school math and science teachers, respectively, are teaching out of field (Texas Education Agency, 2017). The *teach*HOUSTON program has helped to narrow this qualification gap over the past decade.

Fig. 5 shows the percentage of graduates (347) by major since the inception of *teach*HOUSTON. Nearly half (45%) earned math degrees and 27% earned biology degrees. Although degrees earned in chemistry and physics are 6% and 4%, respectively, this makes a significant contribution to areas where teachers are in high demand. In the majors with lower graduation numbers, i.e., computer science, efforts are being made to increase enrollment and put in place courses/ programming to attract and retain students in these areas. This is of great importance since The White House (2016) recognizes computer science as a "new

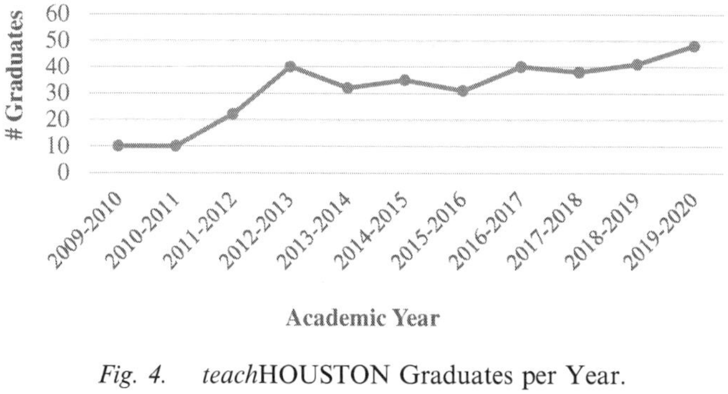

Fig. 4. *teach*HOUSTON Graduates per Year.

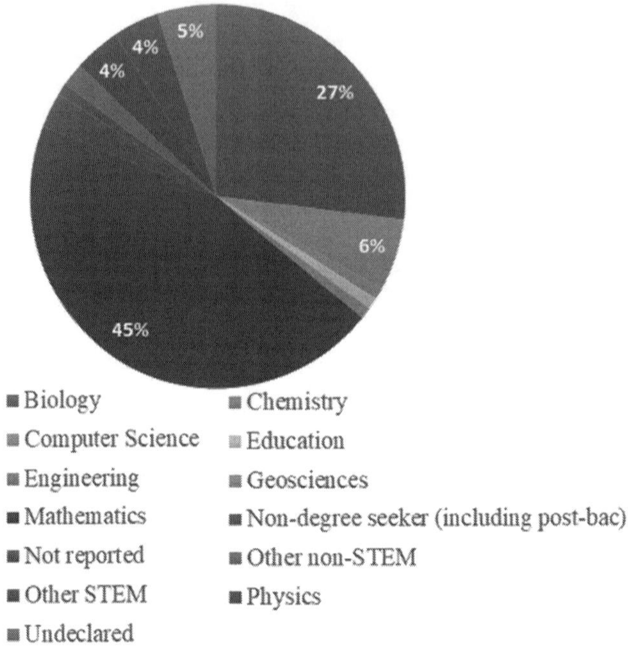

Fig. 5. *teach*HOUSTON Graduates by Major.

basic" skill necessary for economic opportunity and social mobility. Needless to say, there is a significant shortage of computer science teachers that is preventing schools across the nation from offering computer science courses (Gal-Ezer & Stephenson, 2010; Menekse, 2015). Through a recently awarded NSF grant, DUE # 1959936, which focuses on recruitment and retention of STEM majors in high-needs fields like computer science, the *teach*HOUSTON program will continue to address shortages in these areas.

Graduates by Gender

Scholars have noted that the faculty's gender is an important factor in female students' completion of STEM degrees and persistence in STEM professions. In a longitudinal study, Bottia, Stearns, Mickelson, Moller, and Valentino (2015) contended that female high school STEM teachers have a powerful effect on girls' decision to major in STEM. They stated that the proportion of female STEM teachers during precollege years has an important influence on girls' STEM outcomes.

At the national level, the gender rates of teachers in public schools are 77% female and 23% male (National Center for Education Statistics, 2018). The *teach*HOUSTON program follows the national trend for STEM teachers in graduating more female (63%) than male (37%) students, as shown in Fig. 6.

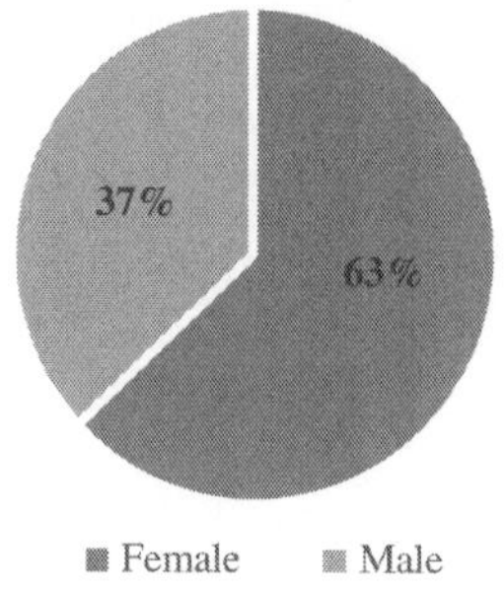

Fig. 6. *teach*HOUSTON Graduates by Gender.

With the underrepresentation of women and girls in the STEM field (Craig, Verma, Stokes, Evans, & Abrol, 2018), *teach*HOUSTON is fostering gender inequity through the production of more female STEM teachers to inspire young female students in pursuing STEM fields.

Graduates by Race/Ethnicity

Empirical literature contends that minority student's academic motivation and outcome improve when they are taught by minority teachers due to similarity in life experiences and cultural backgrounds (Boser, 2014; Dee, 2005; Pitts, 2007). Diversity also benefits white students as interactions with students from different backgrounds help foster a sense of community and build social trust (Boser, 2014). Over the past decade, *teach*HOUSTON has produced a highly diverse pool of STEM teachers who inspire underrepresented students to enter into STEM-related majors and professions. As shown in Fig. 7, Hispanics make up the largest percentage of graduates with 33%, followed by Asians with 30%, whites with 25%, and Blacks/African Americans with 8%. The modest percentage of Black/African American graduates may be due to difficulties in recruiting Black/African American students in the program, given the low percentage of 6.8% for the 2019-2020 academic year enrolled in the College of Natural Sciences and Mathematics at the University of Houston. As the United States becomes more diverse in rational and ethnic backgrounds, it is crucial that *teach*HOUSTON continues to reinforce its approach to recruit, prepare, and retain a diverse group of STEM teachers in the future.

Time to Degree Completion

One of the great aspects of the *teach*HOUSTON program is the specialized degree plans which will allow students to graduate with their major degree and teacher certification in 4 years. Out of the 347 graduates, 162 started as first time in college (FTIC) students, i.e., those that start as freshmen. On average, FTIC students completed the program in four years, which is shown in Fig. 8. Researchers report that female and minority students are less likely to persist in

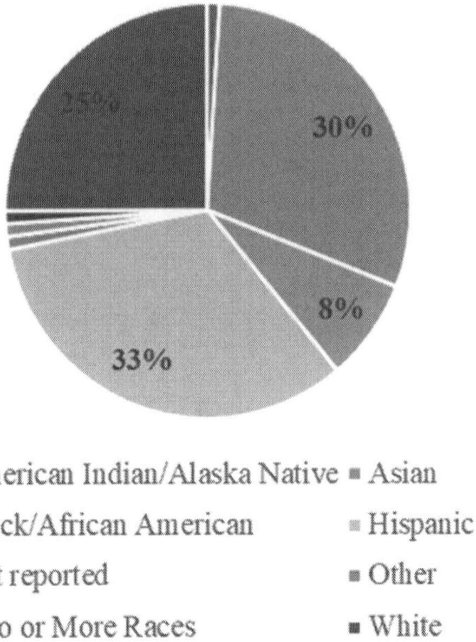

Fig. 7. *teach*HOUSTON Graduates by Race/Ethnicity.

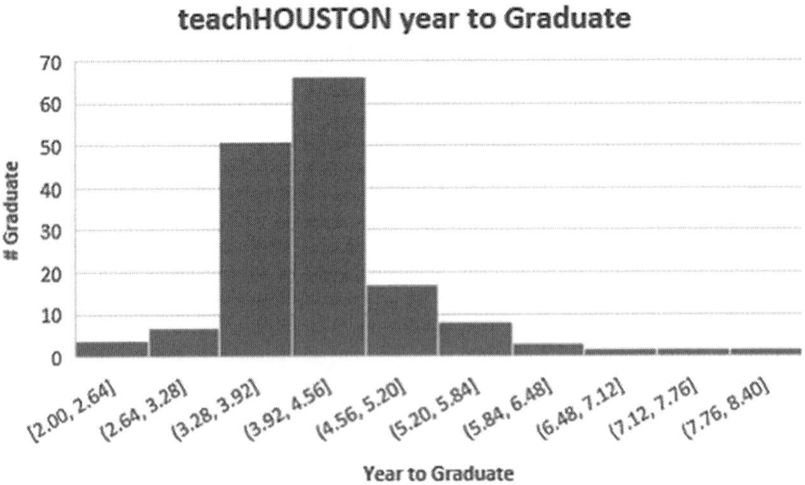

Fig. 8. *teach*HOUSTON Years to Graduate.

the STEM fields than their male peers (Mau, 2016; Toven-Lindsey, Levis-Fitzgerald, Barber, & Hasson, 2015). However, this is not true in the *teach*-HOUSTON program as FTIC female students graduate, on average, in 4.02 years while their FTIC male counterparts graduate in 4.24 years (Fig. 9).

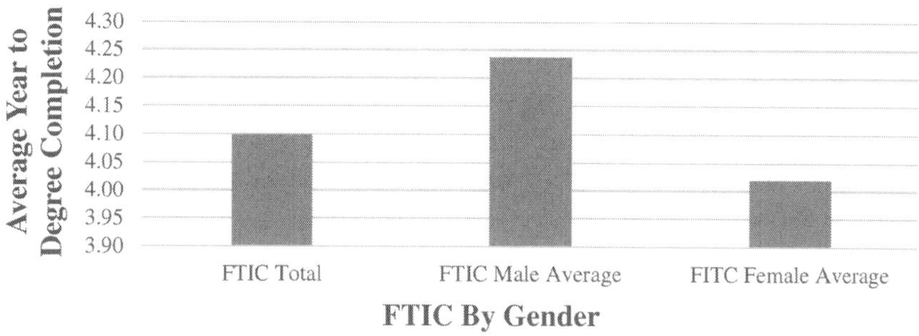

Fig. 9. First time in college (FTIC) students' Average Year to Degree Completion by Gender.

Retention Rate

This critical shortage of STEM teachers is exacerbated by problems with teacher retention which is prevalent in the Houston area where 50% of novice teachers (having five years or less of service) quit teaching by their fourth year of service (National Public Radio, 2014). Moreover, teachers from the leading post-baccalaureate program maintain an average retention of only 2.1 years. Teacher attrition places an insurmountable burden on our community and state with an annual cost of $800 million to the Texas economy (Craig, 2014). As such, retaining teachers in the classroom is critical for providing quality STEM education to K-12 students. The *teach*HOUSTON graduates' five-year retention in the classroom is an impressive 87%, which is higher than other Texas certification pathways' retention rate, i.e., the University Graduate Program (75%), University Post-Baccalaureate Program (68%), and Alternative Certification Program (66%) (Texas Education Agency, 2018).

During 2009–2014, *teach*HOUSTON produced 132 new STEM teachers, of which 115 teachers (or 87%) are currently in their fifth year of teaching or above. Table 1 represents the number of years of teaching experience for 314 *teach*HOUSTON graduates (Spring 2020 graduates are not reported). The "*teach*HOUSTON Program Year of Graduation" includes both spring and fall semesters and "Years of Teaching Completed" reflects the number of completed years teaching in the field. For example, those in the "4 years of experience" column are currently teaching their fifth year. Note that students graduating in a fall semester are about a year behind in experience compared to those who graduated in the spring of the same year. This is reflected in the table by having groups from the same year with one less or one more year of experience than the other.

Where Are They Now?

As of late 2019, 86.6% (or 272/314) of the *teach*HOUSTON graduates are actively teaching in a secondary school or continue to work in or pursue higher education and 13% left the education sector and work in non–education-related careers. Table 2 gives a breakdown of the current careers of the graduates.

Table 1. *teach*HOUSTON Graduates' Years of Teaching Experience.

*teach*HOUSTON Program Year of Graduation	Years of Teaching Completed										
	<1	1	2	3	4	5	6	7	8	>9	Total
2019	41										41
2018		37									37
2017		26	11								37
2016		12	21	1							34
2015		1	12	20							33
2014		1		14	25						40
2013				2	9	20					31
2012					1	8	22	2			33
2011							1	12			13
2010								4	7	1	12
2009								1		2	3

Table 2. Current Careers of *teach*HOUSTON Graduates.

Where Are They Now?												
	Year of Graduation											
Current Career	2009	2010	2011	2012	2013	2014	2015	2016	2017	2018	2019	Grand Total
Continued their education		1		2	1	1	3	1	1	3	2	15
Currently teaching	1	5	12	24	25	37	27	30	30	30	36	257
Non-education career	2	6	1	7	5	2	3	3	6	4	3	42
Grand total	**3**	**12**	**13**	**33**	**31**	**40**	**33**	**34**	**37**	**37**	**41**	**314**

Of the graduates who are currently teaching in a secondary school, 88% (225/257) are teaching in school districts in the Greater Houston area (see Fig. 10) including the high-need school districts, Houston ISD, Pasadena ISD, Fort Bend ISD, Katy ISD, and Aldine ISD. Through these positions, the *teach*HOUSTON graduates are transforming STEM education for thousands of students of low socioeconomic status in diversely populated areas.

Major Findings

The *teach*HOUSTON program has experienced success in recruitment, preparation, and retention of secondary STEM teachers, most of which teach in Greater Houston. Three areas we especially spotlight are the increased numbers of physics teachers, the diversity of their graduates compared to the state of Texas

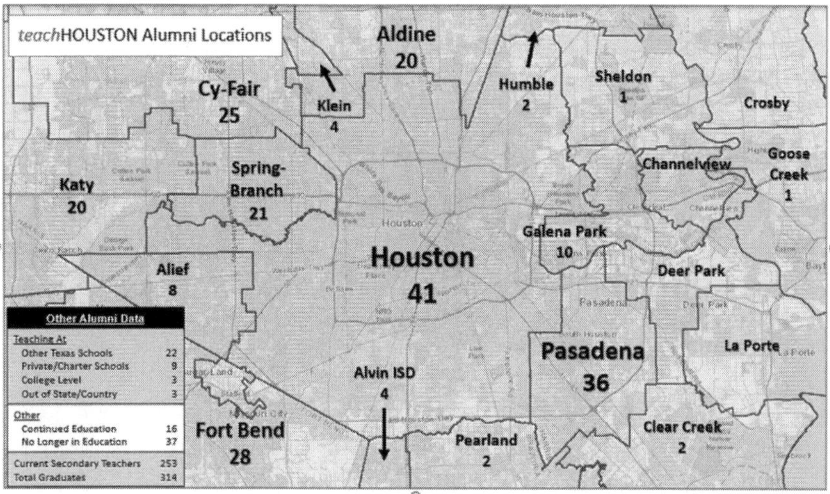

Fig. 10. Location and Number of *teach*HOUSTON Graduate Teaching in School Districts in the Greater Houston Area.

and the United States, and the creation of a culture of teacher leadership. This next section will highlight these areas and will include the journeys of three teacher leaders.

Physics Teacher Production

One of the main goals of now *teach*HOUSTON Associate Director, Paige Evans, was to increase the number of physics teachers in the *teach*HOUSTON program as there was a drastic shortage of physics teachers both locally and nationally. Because of her 17 years of experience in teaching physics in Houston area public schools coupled with the extreme shortage of physics teachers in the Houston area, she naturally gravitated toward building a collaborative relationship with the Physics Department in order to increase the potential of developing physics teachers.

She quickly learned that Donna W. Stokes, professor and undergraduate chair in the Physics Department, and now the Associate Dean of the College of Sciences and Mathematics, shared her vision. Subsequently, they strategized and formed a plan to develop future physics teachers. To this end, they developed a physics minor specifically for math majors who were pursuing teacher certification. Additionally, they created specialized degree plans and obtained approval for teacher certification from the state for physics majors and minors without adding any time to their degree plan.

In 2012, they, Stokes as PI and Evans as co-PI, applied for and were awarded a grant through the NSF, DUE # 1240083, for the Robert Noyce Teacher Scholarship Program that aimed to encourage STEM majors and professionals to become K-12 teachers to serve in high-need schools. The UH Robert Noyce scholarship Program addressed recruitment, preparation, and retention of science majors for secondary education in physics and chemistry. The goals of the program not only included awarding scholarships but also offered summer internship for early exposure to formal and informal teaching experiences (Stokes, Evans, & Craig, 2017), developing and enacting a "Physics by Inquiry" course and evaluating instructional strategies for training highly qualified physics and chemistry teachers.

The "Physics by Inquiry," PHYS 4342, was developed to engage future science teachers in interactive, inquiry-based teaching pedagogies for physics and chemistry resulting in a larger number of teachers prepared to effectively teach physics and/or chemistry at the 7-12 level. Due to the success of the Physics by Inquiry course, faculty created a similar physics course for future middle school science teachers, Physics for Preservice Teachers (PHYS 4345), to teach middle school physics concepts while incorporating effective teaching strategies. Prior to the inception of this course, the middle school teacher candidates lacked adequate content acquisition with physics-related topics which are required as part of Texas' grades 6–8 educational standards. The course not only introduces preservice middle school science teachers to physics content but also provides experiential learning with hands-on methods for teaching the content. Upon completion of the course, teacher candidates are able to enter the classroom with tools, resources, and expertise in an otherwise neglected area of middle school science.

These courses generated through collaborative efforts have been integral in increasing the number of physics teachers for the Houston area. As mentioned, prior to the *teach*HOUSTON and Physics Department collaboration, there were zero graduates certified to teach physics in the previous decade. Since the Fall 2014, *teach*HOUSTON has produced an average of three certified physics teachers per year and three of these teachers received Teacher of the Year honors for their school. Table 3 shows these results.

Other areas of collaboration include working with physics faculty on the STEM Teaching Equity Project, a professional development program for in-service physics teachers, the high school physics contest, and the Mars Rover Competition. Both *teach*HOUSTON and Physics faculty are continuing this collaboration with a new grant NSF grant, DUE #1950036, Advancing Cultural and Computational Engagement in STEM Scholars (UH-ACCESS), which will be explained in more detail below.

Table 3. Graduation Numbers for Certified Physics Teacher 2014–2019 by Academic Year.

Graduates	2014–15	2015–16	2016–17	2017–18	2018–19
Physics teacher education graduates[a]	3	2	2	4	2

[a]Any student who graduated with a major or minor in physics and completed the *teach*HOUSTON teacher certification program.

STEM Teacher Workforce Diversity

It is no secret that the US student population is becoming increasingly diverse. This trend is expected to continue, and it is predicted that in the next two decades, a majority of students will belong to race-ethnic minorities (Schmitz, Nourse, & Ross, 2012). Still, the demographics of teacher candidates in the United States do not reflect the cultural, racial, and linguistic diversity of the student population that they will be teaching (McNulty & Brown, 2009). This disparity is especially concerning provided the benefits in student learning that are attained when teachers from minority populations teach minority students (Basit & McNamara, 2004). Moreover, scholars have argued that in order to facilitate change, a critical mass of minority teachers must enter the field and be present in school environments in order to effectively advocate for the academic needs of minority students (Atkins, Fertig, & Wilkins, 2014). Ladson-Billings (2005) spoke of the racial and cultural mismatch between students and their teachers by explaining that the gap in diversity poses an additional challenge for teachers to be successful with K-12 students of color. McNulty and Brown (2009) affirmed this notion by suggesting the benefits associated with minority teachers such as their ability to cultivate a sense of community and belonging for their students. To meet this challenge, teacher preparation programs across the country have adopted innovative methods for the recruitment of teacher candidates from racial and ethnic diverse backgrounds (Schmitz et al., 2012). Accordingly, the *teach*HOUSTON program has been charged with this mission, since the inception of the program, and continually produces teachers that reflect the diversity of the city and student population. *teach*HOUSTON graduates are 34% Hispanic, 30% Asian, 25% white, 8% African American, and 3% other as compared to US and Texas teachers (Texas Education Agency, 2018; NCES Indicator, 2018) as shown in Fig. 11. As noted earlier, the production of diverse teachers is critical given that successful learning outcomes occur when minority teachers serve as role models and reflect the student population (Basit & McNamara, 2004; Dee, 2005; Pitts, 2007). Moreover, the population mix in Texas will resemble that of the United States in 2040 (Walsh, 2013). Hence, it is essential that teacher preparation produces teachers that reflect the diversity of their students and can develop lessons that celebrate their cultural and linguistic backgrounds.

*teach*HOUSTON *Teacher Leaders*

Defining Teacher Leadership

As established, teachers are the single most critical factor when it comes to improving student learning outcomes (National Commission on Teaching and America's Future, 1996). Hence, it is no surprise that, over the decades, efforts in school reform have emphasized the importance of teacher quality and development. Although the notion of teacher leadership is often mentioned, rarely do authors define the term (York-Barr & Duke, 2004). In their extensive literature review of teacher leadership, York-Barr and Duke (2004) recognize

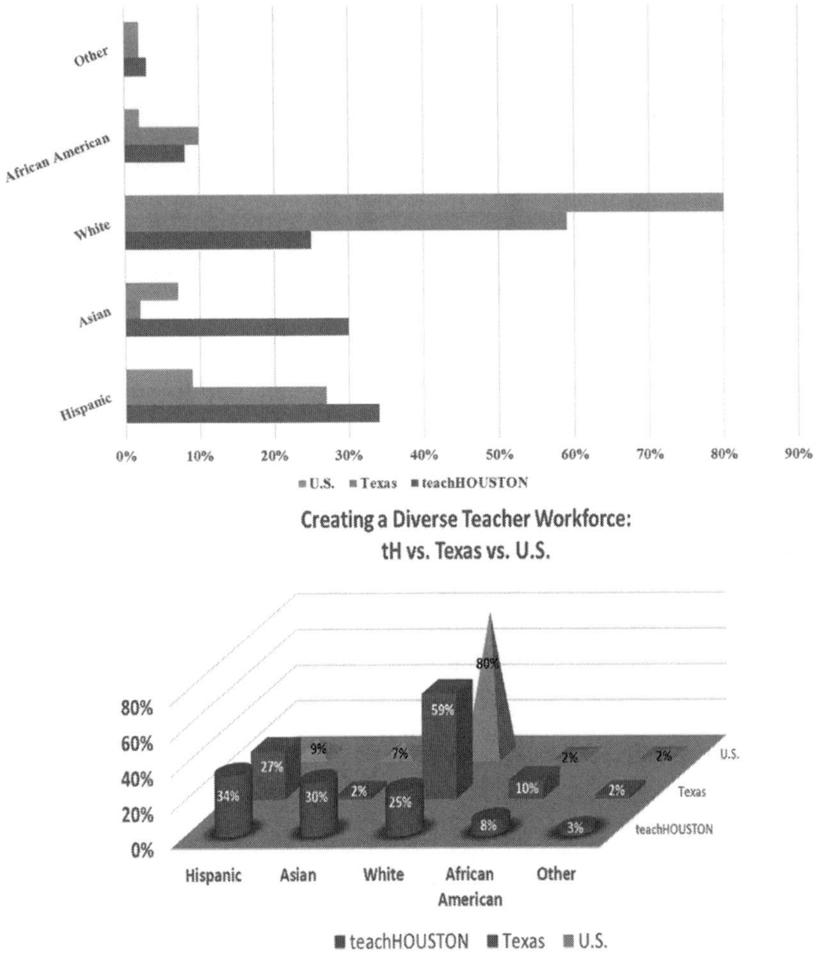

Fig. 11. Creating a Diverse Teacher Workforce: *teach*HOUSTON versus Texas versus United States.

three strands of teacher leadership that have been described by scholars, the first of which includes administrative roles such as department chair or content team lead. These positions do not change the practices of a school. Rather, they reinforce the administration's agenda. The second strand of teacher leadership includes instructional agency and is accomplished through roles such as curriculum writer and mentor teacher. The third strand is the one in which the teacher is viewed as an individual with agency to transform the school's culture and drive the administrative agenda. For the purpose of this chapter, teacher leadership is defined through the words of Childs-Bowen, Moller, and Scrivner (2000):

> We believe teachers are leaders when they function in professional learning communities to affect student learning; contribute to school improvement; inspire excellence in practice; and empower stakeholders to participate in educational improvement. (p. 28).

Over the past decade, *teach*HOUSTON has produced 347 quality secondary STEM teachers. Many are currently serving in leadership positions for their school districts. These positions vary between roles inside and outside of the classroom. For example, some *teach*HOUSTON alumni are serving as assistant principals and instructional coaches, while others are department chairs, team leads, and curriculum writers for their school districts. All of these positions allow graduates to extend their impact toward colleagues and other fellow STEM educators.

*teach*HOUSTON alumni also utilize their leadership capacity to mold extra-curricular opportunities for students through sponsorship of organizations and clubs. Moreover, graduates are actively investing in their development and surveys show that they are members of professional learning organizations. For example, *teach*HOUSTON alumni and Advanced Placement Physics instructor, Jane Li, is an Ambassador for the STEP UP program, a national program that aims to help engage young women into Physics and change deep-seated cultural views about physicists. She is also a recipient of the prestigious T.E.A.C.H. (Texas Educators Achieving Classroom Honors) award. In fact, many *teach*HOUSTON graduates are recipients of awards such as Teacher of the Year and Rookie Teacher of the Year.

It is clear that *teach*HOUSTON graduates are not only staying in public education but are also cultivating a legacy for the program that is grounded in their aspirations to improve the state of STEM education. In the forthcoming section, three stories of *teach*HOUSTON graduates who are serving in leadership positions are presented. The first story is about a graduate who is now an instructional coach, the second is about a classroom teacher who is leading the way in his content area, and the last story features a narrative from one of the authors who is a master teacher for the program and also an alumni.

Administrative Leadership: Instructional Coach

Brianna vividly recalls the meeting that changed her career trajectory and yielded a realization of her true calling. Upon her mother finding an article about *teach*HOUSTON in the local newspaper, Brianna was intrigued and scheduled an advising session. Within an hour of speaking with the *teach*HOUSTON advisor, Brianna knew that the path she had previously selected would not be the one she would pursue. Instead, Brianna switched her degree plan to include *teach*HOUSTON and remained confident that in that moment she made the right decision.

When asked about how *teach*HOUSTON prepared her for the classroom, Brianna explains that the program instilled in her a sense of organization and purpose. "*teach*HOUSTON allows students to become master planners before entering their own classrooms. Students from *teach*HOUSTON are creative, insightful, and meticulous." Brianna recalls all of the carefully planned procedures she implemented during her first year, such as name tents and color-coded binders for each student. Although, at the time, her colleagues warned that her efforts were time wasted on students who would continue to underperform, Brianna, based on her teacher preparation background, knew different. Hence, by the end of that first year, she was receiving well-deserved praise from school and district leaders because her students had the highest levels of growth and accomplishment on their Algebra I state exam scores.

Brianna's success in the classroom presented her opportunities to extend her influence across the department, school, and district. She served as a department teacher and a skills specialist and is currently taking on the role of instructional coach for a secondary mathematics department. Brianna attributes her leadership capacity to lessons learned during her time with *teach*HOUSTON.

> One of the most important aspects of leadership are systems and procedures. A leader must have a vision. They must see systems and procedures and work toward successfully fulfilling the needs. *teach*HOUSTON was extremely influential in allowing me to do this.

Brianna has come a long way since that initial meeting with the *teach*-HOUSTON advisor. She is currently working on her doctorate in educational leadership and continues to stay involved with *teach*HOUSTON through events that support the development of students and new teachers from the program, along with conference presentations with master teachers and fellow alumni.

Classroom Teacher Leader

Saul remembers sitting in his first *teach*HOUSTON course and feeling excited about the journey ahead. *teach*HOUSTON prepared him to face the challenges of a first-year teacher and then supported him along the way. "My first year of teaching was a major test to all of the practices I had learned during my time in *teach*HOUSTON. I successfully passed the test." According to Saul, the obstacles encountered during the first year only helped solidify his commitment toward teaching. He also gained a sense of clarity regarding his vision for the future. Saul was determined to have a positive impact in the lives of as many students as possible and to do so he realized that he would need to share his skills and passion with other teachers.

Saul's second year of teaching was filled with reflections about his practice, philosophy of teaching, and desire to mentor. *teach*HOUSTON helped cultivate opportunities for Saul to build his mentorship skills. Saul began serving as a mentor teacher for preservice teachers in the program. Saul viewed the role as "a great opportunity to push teachers to become their very best." Eager to share his ideas, Saul also began providing professional development for teachers in his school district. He felt empowered to present in front of other educators due to the confidence gained from conference opportunities provided through his *teach*HOUSTON network. Saul worked with master teachers and fellow graduates to design professional development sessions centered on research-driven teaching practices.

Currently, Saul is serving as the Lead Teacher for the Newly Arrived Center at his campus where he teaches ESL/ELL (English as a Second Language/English Language Learners) Biology classes. Saul is also a cooperating teacher for the *teach*HOUSTON program, which means that he is responsible for mentoring and guiding a student teacher. Cooperating teachers model effective teaching, observe their student teacher, give advice, and provide feedback for their student teacher, as well as assist with planning and co-teaching. Saul regularly presents for his district and around the nation through *teach*HOUSTON. Moreover, he is often invited to serve on the *teach*HOUSTON alumni panel where undergraduate students in the preservice teacher preparation program have the opportunity to interact with and learn from alumni. Saul's journey with *teach*HOUSTON began the moment he learned about the first course in the program and continues to this day as he now embodies the role of mentor to aspiring STEM teachers.

Teacher Leader to Master Teacher

This section narrates the path of coauthor, Mariam Manuel, as she reflects on her journey to becoming a STEM teacher, an instructional coach, and then a faculty member in the *teach*HOUSTON program. Mariam shares her story in her own words:

> For as long as I can remember, I have always aspired to teach. However, the pathway to the classroom was not always clear. Growing up I felt a pressure that many students experience when their parents interpret their love for science as an assurance for journeying into a medical related profession. In an effort to please my family I subscribed to the belief that I was destined for a career in medicine. I entered the University of Houston as a pre-pharmacy student and was soon faced with the reality I had always known, I did not want to be a pharmacist. It was strikingly obvious considering that I used my study groups as an opportunity to teach rather than to simply study with my classmates. I wanted to be a teacher, but at the time, I did not know how.

Mariam then shared what happened next:

> That changed the day when my calculus professor presented our class with a new route designed to certify STEM majors in secondary STEM teaching. Within 24 hours, I was enrolled in the program that would become the guiding force for my career for many years to come. *teach*HOUSTON provided me a pathway to pursue my aspirations. Even though I had set my mind to becoming a teacher, I did not know the first thing about embodying an inquiry-based approach. Through the course work I learned about strategies, some of which I instinctively knew were good teaching practices, and others that inspired me to grow further in the field. Not only did I enjoy the coursework, but, for the first time in my college experience, I felt a sense of community and belonging. The eight other *teach*HOUSTON students in the first cohort became my close friends. Many of us were also hired as student workers for the program and through that role were able to gain additional mentorship from the master teachers who taught our classes. We were all pursuing STEM degrees while venturing on a pathway to teaching. Our decision to teach upon graduation was one that many of our friends and families dismissed or challenged. This allowed us to bond closer together and support one another. It was not long before we all walked across the graduation stage and entered classrooms of our own. Although our diplomas held tightly within our grasp signified the end of our undergraduate journey, the impact *teach*HOUSTON would have on our careers was far from over.

Mariam then looked backward on what had transpired in her career:

> During my first year in the classroom I faced the common challenges that many first-year teachers experience along with some unexpected obstacles. I was assigned multiple preps and was struggling to collaborate with colleagues who did not share my vision for teaching through inquiry. Additionally, the national recession led many school districts to lay off teachers, especially those in their first three years, commonly known as the probationary period. Hence, the fear of losing the dream career I spent years working towards was constantly looming over my head. However, *teach*HOUSTON continued to serve as my rock and I leaned on them for much needed support. The assistance I received from *teach*HOUSTON came in many forms. I was visited by master teachers who continued to help with curricular resources via email. I also had the camaraderie of my fellow alumni who were experiencing similar situations. We allowed each other to vent and continued to cheer each other along until we all made it to the finish line. The years that followed were easier and the support that was once critical to my survival became my motivation to help others. As such, through partnership with a master teacher and other alumni, we formed the *teach*HOUSTON alumni advisory board. The board allowed us to host networking sessions, develop an online bank of resources, and provide input on programmatic decisions.

Mariam Manuel concluded:

> After serving as a classroom teacher, instructional coach, and the instructional specialist charged with opening a STEM center for the school district, I was invited to interview for the position of master teacher for the *teach*HOUSTON program. The prospect of serving in a role occupied by the individuals who inspired and shaped my development as a teacher was humbling. To this day, I am acutely aware of the responsibilities associated with the position of a master teacher and the opportunity I have to make an impact in the lives of many aspiring STEM teachers. Accordingly, my experiences as a preservice teacher has shaped my philosophy and continues to guide my instructional decisions.

LEADING THROUGH INNOVATION

Grant Opportunities

Through obtaining three recent federal grants, *teach*HOUSTON has continued to evolve and adapt its programming. The innovation produced through these grants aims to meet the needs of the most diverse city in the country which is vital as the present-day population mix is projected to resemble the United States in 2040 (Walsh, 2013). This section discusses each project and the mechanisms through which it supports and sustains teacher development.

National Science Foundation Grant, University of Houston – Leadership through Equity and Advocacy Development (UH-LEAD)

Launched in 2018, the goals of UH-LEAD are to (1) retain STEM teachers in high-need school districts; (2) cultivate teacher leaders in high-needs districts to mentor preservice teachers and work collaboratively with in-service teachers who are focused on implementing the STEM disciplines through culturally responsive teaching in secondary classrooms; and (3) build a cadre of STEM teacher leaders in high-need school districts that are equipped to advocate for change at the local, regional, and national platform. These goals are being accomplished through the following activities:

- Supporting 30 Master Teacher Fellows (MTFs), 15 in cohort 1 and 15 in cohort 2, with tuition and fees of approximately $16,000 through a 14-month Curriculum and Instruction (CUIN) Master's program with an emphasis in STEM education;
- Offering $10,000 annual salary supplements for each fellow for four years subsequent to completing the master's program;
- Providing face-to-face and virtual professional development that includes culturally responsive teaching in STEM, pedagogical content knowledge, and instructional coaching in STEM;
- Developing leadership skills so that MTFs will assume leadership roles in facilitating culturally responsive teaching infused with effective pedagogies in STEM through mentoring both preservice teachers and in-service teachers in their schools/districts; and
- Collaborating with the National Math and Science Initiative (NMSI) to provide MTFs with STEM pedagogical content knowledge courses that MTFs will later infuse with culturally responsive teaching.

The research generated from this project is contributing to literature on preparing and retaining STEM teacher leaders in high-need secondary schools by (1) incorporating culturally responsive pedagogy (CRP) in secondary STEM preparation and professional development; (2) cultivating MTFs leadership skills through instructional coaching; and (3) developing and evaluating MTFs' abilities to be change agents in professional development and CRP in their schools/districts.

This grant project is progressing extremely well with 30 MTFs around the Greater Houston area embarking on the transformation of STEM public education. Fig. 12 shows school placements of MTFs. Over the past three years, MTFs have earned their master's degrees in STEM education and have begun receiving professional development in areas such as instructional coaching. Their professional development is designed strategically to support the areas of need expressed by the fellows. Currently, they are building their capacity as professional development providers for school districts and conferences. Through these efforts, MTFs will extend their impact beyond the classroom and deliver professional development from grant-related topics/goals to other interested STEM educators.

Department of Education Grant, ACP for STEM at UH

During the fall of 2019, *teach*HOUSTON piloted a uniquely designed alternative certification program, Alternative Certification Pathway for STEM (ACP for STEM at UH), with a Department of Education grant. The overarching goal of this initiative is to create a pipeline of highly qualified secondary STEM teachers via this new approach to alternative certification involving substantial field-based

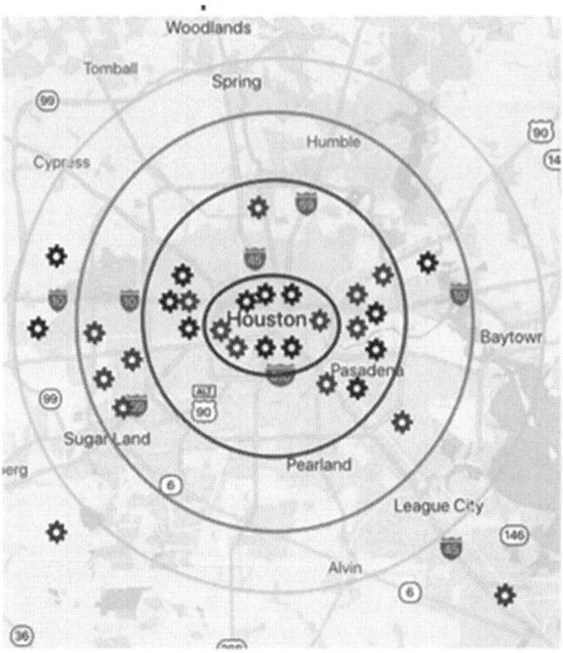

Fig. 12. Schools Placement of Master Teacher Fellows (MTFs).

experience, emphasizing promising teaching practices, and innovation in STEM instruction via the use of CRP. This program targets those who already have a STEM degree (including career changers) or those who have only a few courses left that are seeking secondary STEM teacher certification. In this new program, participants receive the expertise, quality, and support provided from the *teach*HOUSTON program through the teacher certification plan.

ACP for STEM at UH is a unique program that includes essential components from traditional teaching certification programs such as quality of feedback and coursework, field experiences, mentoring, and support. However, unlike a traditional pathway which generally takes place over the course of three to four years, ACP for STEM at UH is completed within nine months allowing candidates to earn their certification in an accelerated manner and enter the classroom with support and resources.

Students complete the program in nine months and receive certification in Mathematics 7–12, Science 7–12, Physics/Mathematics 7–12, Life Science 7–12, or Chemistry 7–12. Students complete the same coursework as in the traditional *teach*HOUSTON program; however, it is enhanced with the inclusion of CRP. The instruction model is daytime, face-to-face classes, hybrid classes, and field experiences in local public schools. Curriculum emphasizes the following:

- Student-centered lesson planning in STEM including inquiry-based learning
- Classroom management
- CRP
- Using technology in meaningful ways with students
- Developing twenty-first century skills
- Over 500 hours of authentic field-based experiences in local public schools, including a full semester of student teaching.

During the first semester, students meet as a cohort on Tuesdays and Thursdays and complete all the field-based courses, with the exception of student teaching which takes place during the second semester. The courses are modified from the traditional *teach*HOUSTON pathway needed for teacher certification. These field-based courses include the following:

(1) Introduction to STEM Teaching: 3 Credit Hours
(2) Secondary STEM Methods or Inquiry-Based Methods (i.e., Classroom Interactions) Hybrid: 3 Credit Hours
(3) Secondary STEM Capstone or Advanced Inquiry-Based Methods (Multiple Teaching Strategies) Hybrid: 3 Credit Hours

Additionally, the students take a *teach*HOUSTON course, *Education in a Multicultural Society* (EMS) as a 3-credit hour hybrid course. The curriculum for EMS is purposefully designed to explore dimensions of knowledge and knowledge construction, institutional and instructional inequities, the role of personal bias, and the way these factors impact STEM teaching. Students examine the meaning of their cultural paradigm (and that of their students) and its impact on the STEM learning environment along with current understandings of culturally responsive pedagogies. By building awareness of a variety of perspectives, students will better understand how to maximize achievement and create a more equitable educational STEM learning environment for all students. Instructors purposefully vertically align all coursework to embed the implementation of culturally responsive instruction with inquiry-based STEM instruction. On Monday and Wednesday, they observe and/or teach for at least six hours each week. During the second semester, students complete student teaching, which is a full-time 14-week, all-day immersion experience into public school settings.

Although the courses are listed as four separate courses (12 credit hours) for the students, *teach*HOUSTON has seamlessly turned these courses into one continuous three-hour course that takes place on Tuesdays and Thursdays. These courses are reinforced by their field experience on Monday and Wednesday where students spend at least six hours in the field observing and teaching with expert mentor teachers in high-need school districts. The instructional team reconfigured the curriculum to cover all essential objectives of the courses with the integration of CRP. The traditional pathway is designed to include redundancy, which helps to ensure the thorough preparation of undergraduates who take the courses over the span of four years often taking a semester off in between. However, the ACP for STEM at UH grant program takes place in a shorter amount of time with a more concentrated focus on the topics. As such, the accelerated sequence still provides adequate time to authentically prepare teacher candidates for their student teaching semester. In fact, the faculty/master teachers, alumni, and program advisor affiliated with traditional *teach*HOUSTON pathway who are now working the ACP students firmly believe that these candidates will be *thoroughly* prepared to enter the classroom and hold certain advantages. For instance, the ACP candidates experience an immersion into teaching and are not distracted by the responsibility of also completing rigorous STEM content coursework. Moreover, the cohort model provides a solid system of support and camaraderie between the candidates. Also, ACP candidates are able to work with two consistent faculty members/master teachers which helps create a bond of trust and mentorship.

The program has been successful in its first year and graduated 17 highly qualified STEM teachers of which 9 are females and 7 are males. Eleven of the candidates are career changers and completed the program as post-baccalaureates. The other six students in the cohort either had completed their STEM degrees or had only a couple of courses left in their degree plans. Eight received a 7-12 composite science teaching certificate, eight received a 7-12 mathematics teaching certificate, and one received a 7-12 math/physics teaching certificate. All will be teaching in Greater Houston during Fall 2020. We anticipate that cohort 2 will have approximately 22 STEM teacher candidates with the majority being STEM career changers.

National Science Foundation Grant, Advancing Cultural and Computational Engagement in STEM Scholars (UH-ACCESS)

Recently awarded on April 1, 2020, UH-ACCESS will create a pipeline of secondary STEM teachers trained in high-need fields including computer science, technology, and physics by building a new partnership with the College of Technology and expanding partnerships with the Departments of Computer Science (CS) and Physics with *teach*HOUSTON. The main goal of UH-ACCESS is to prepare 30 STEM majors to become effective secondary STEM teachers in high-need school districts to address the growing demand for producing a workforce with skills related to computer science and technology. Objectives include the following: (1) recruit STEM majors in CS, technology, and physics into *teach*HOUSTON, (2) support these students by offering $12,000 per year scholarships, (3) use a "culturally responsive" selection criteria to obtain scholars demonstrating a dedication to social justice and exhibiting sociopolitical awareness needed for teaching in high-need school districts, (4) develop/implement an inquiry-based computer science education course for those majoring/minoring in CS as well as those majoring in STEM fields who wish to incorporate CS into their STEM lessons, and (5) establish a Teacher Interest Group (TIG) to build community, offer support, and provide opportunities to participate in community events incorporating a culturally responsive context to help scholars teach underrepresented student populations and connect to their communities. This will develop an infrastructure, which creates a pipeline for preparing teachers in CS, technology, and physics, who can persist in high-need school districts.

The components of the UH-ACCESS program will expose preservice teachers who already exhibit a sociocultural awareness to diverse environments early via ongoing field experiences and through community events in the TIG. A main component of the TIG will be to better prepare preservice teachers to be effective in high-need school districts by incorporating CRP into their STEM classroom curricula and cocurricular activities as current students and teachers reported that they need additional support integrating CRP into STEM lessons and in the management of their classrooms. For example, the TIG will introduce preservice teachers to community leaders and activities which will heighten their awareness of how cultural considerations in STEM education can impact the students they teach as research confirms that CRP is a vehicle for increasing student success in high-need classrooms (Howell, Lewis, & Carter, 2011; Perry & Delpit, 1998). Moreover, the TIG will provide educational experiences that help preservice teachers build CRP teaching practices into their curriculum and help them make connections between the curriculum and equity/social justice issues in their communities. Additionally, through these purposeful community-focused interactions, project work, and personal connections to individuals involved in social justice advocacy, scholars will become exposed to and engaged in community-based learning contexts that inform and improve their practice of culturally inclusive educational pedagogies as future STEM educators (Mckenna, Greene, & Burke, 2017). Preservice teacher scholars will receive mentoring and support for adopting and implementing CRP into their STEM classes from MTFs of the aforementioned UH-LEAD grant who are particularly prepared to teach in Houston's urban classrooms. In addition, scholars graduating from UH-ACCESS will be supported for the first three years of their careers through an induction program.

LOOKING AHEAD

Although *teach*HOUSTON has an excellent record of producing highly qualified STEM teachers, the need across the nation far exceeds what cumulatively is being done. For this reason, *teach*HOUSTON strives to foster a sense of leadership and change agency within its graduates. They are able to take the practical knowledge cultivated during their years with the program and share this with teachers in schools, districts, and around the country through conference presentations and professional development. Moreover, the leadership roles held by alumni serve as platforms for them to inform and shape their school's agenda. This chapter has illustrated the program's productivity with respect to a diverse STEM teacher workforce and then spotlighted three alumni and their contributions. Where they are currently and how the *teach*HOUSTON program molded teachers as leaders that are sustained in the teaching profession was shown. Their narratives exemplify *teach*HOUSTON's role in preparing teachers who will be great teachers and leaders in the field. By looking at how alumni are making significant contributions to the teaching profession, it is evident that the *teach*HOUSTON program is working toward reducing the shortage of quality STEM teachers. Just as the graduates continue to find new ways to inform and educate, so does *teach*HOUSTON. By continuing to pursue innovative ways to grow and improve through recent grant projects, the program will further its production of effective STEM teachers and build its efforts to provide quality in-service teacher professional development.

A vital lesson learned in education reform is that "no education system can do better than the quality of its teachers" (Barber & Mourshed, 2007; Department of Education, 2010). The road ahead for *teach*HOUSTON is filled with possibilities of ventures that are yet to be planned and enacted. The master teachers, graduates, and students may appear to be fighting an uphill battle as staggering statistics about STEM education are made public, however, they are doing so with passion, zeal, and determination because at the heart of *teach*HOUSTON lies the belief that teachers can – and do – make a difference as this chapter and other chapters in this book show.

AUTHORS' NOTE

This research was funded by the following grants:

- National Science Foundation DUE Award 1240083, *Recruitment, Preparation and Retention of STEM Students as High School Teachers*
- National Science Foundation DUE Award 1557309, *University of Houston: Learning through Informal and Formal Experiences* (UH-LIFE)
- National Science Foundation DUE Award 1557273, *Collaborative Research: Understanding Robert Noyce Teacher Scholarship Outcomes in Texas*
- National Science Foundation DUE Award 1759454, *Leadership through Equity and Advocacy Development* (UH-LEAD)
- National Science Foundation DUE Award 1950036, *Preparing STEM Teachers through Advancing Cultural and Computational Engagement in STEM Scholars* (UH-ACCESS)
- Department of Education Supporting Effective Educator Development (SEED) Grant, U423A180139, *Expanding and Strengthening the STEM Teacher Workforce Through UTeach*

REFERENCES

Atkins, D., Fertig, A., & Wilkins, V. (2014). Connectedness and expectations: How minority teachers can improve educational outcomes for minority students. *Public Management Review*, *16*(4), 503–526.

Barber, M., & Mourshed, M. (2007). *How the world's best-performing systems count out on top*. London: McKinsey & Company.

Basit, T. N., & McNamara, O. (2004). Equal opportunity or affirmative action? The induction of minority ethnic teachers. *Journal of Education for Teaching: International Research and Pedagogy*, *30*(2), 97–115.

Boser, U. (2014). Teacher diversity report: A new state-by-state analysis. Retrieved from https://cdn.americanprogress.org/wp-content/uploads/2014/05/TeacherDiversity.pdf

Bottia, M. C., Stearns, E., Mickelson, R. A., Moller, S., & Valentino, L. (2015). Growing the roots of STEM majors: Female math and science high school faculty and the participation of students in STEM. *Economics of Education Review*, *45*, 14–27.

Childs-Bowen, D., Moller, G., & Scrivner, J. (2000). Principals: Leaders of leaders. *NASSP Bulletin*, *84*(616), 27–34.

Craig, C. (2014). From stories of staying to stories of leaving: A US beginning teacher's experience. *Journal of Curriculum Studies*, *46*(1), 81–115.

Craig, C., Evans, P., Stokes, D., & Bott, S. (2017). Attracting, preparing and retaining teachers in high need areas: A science as inquiry model of teacher education. In M. Peters, B. Cowie, & I. Mentor (Eds.), *A companion to research in teacher education*. New York, NY: Springer Publishing.

Craig, C. J., Verma, R., Stokes, D., Evans, P., & Abrol, B. (2018). The influence of parents on undergraduate and graduate students' entering the STEM disciplines and STEM careers. *International Journal of Science Education*, *40*(6), 621–643.

Darling-Hammond, L. (2006). Constructing 21st-century teacher education. *Journal of Teacher Education*, *57*(3), 300–314.

Dee, T. S. (2005). A teacher like me: Does race, ethnicity, or gender matter? *The American Economic Review*, *95*(2), 158–165.

Department for Education. (2010). *The importance of teaching: The schools white paper 2010*. London: Department of Education.

Gal-Ezer, J., & Stephenson, C. (2010). Computer science teacher preparation is critical. *ACM Inroads*, *1*(1), 61–66.

Howell, L. L., Lewis, C. W., & Carter, N. (Eds.). (2011). *Yes we can! Improving urban schools through innovative educational reform*. Charlotte, NC: Information Publishing.

Ladson-Billings, G. (2005). Reading, writing, and race: Literacy practices of teachers in diverse classrooms. In T. McCarty (Ed.), *Language, literacy, and power in schooling* (pp. 133–150). Mahwah, NJ: Erlbaum.

Marder, M., & Hamrock, C. (2016). Math and science outcomes for students of teachers from standard and alternative pathways in Texas. *Science*, *126*, 27.

Mau, W. C. J. (2016). Characteristics of US students that pursued a STEM major and factors that predicted their persistence in degree completion. *Universal Journal of Educational Research*, *4*(6), 1495–1500.

Mckenna, M., Greene, S., & Burke, K. (2017). Teaching from the outside in: Community-based pedagogy's potential for transformation. *Transformations: The Journal of Inclusive Scholarship and Pedagogy*, *27*(2), 241. doi:10.5325/trajincschped.27.2.0241

McNulty, C. P., & Brown, M. S. (2009). Help wanted: Seeking the critical confluence of minorities in teaching. *Childhood Education*, *85*(3), 179–181.

Menekse, M. (2015). Computer science teacher professional development in the United States: A review of studies published between 2004 and 2014. *Computer Science Education*, *25*(4), 325–350. doi:10.1080/08993408.2015.1111645

National Center for Education Statistics. (2018). Characteristics of public school teachers. Retrieved from https://nces.ed.gov/programs/coe/pdf/Indicator_CLR/coe_clr_2018_05.pdf

National Commission on Teaching & America's Future (US). (1996). *What matters most: Teaching for America's future: Report of the national commission on teaching & America's future*. New York, NY: The Commission.

National Public Radio. (2014, July 18). The teacher dropout crisis. Retrievd from http://www.npr.org/sections/ed/2014/07/18/332343240/the-teacher-dropout-crisis

Organization for Economic Cooperation and Development. (2005). *Teachers matter: Attracting, developing and retaining effective teachers*. Paris: OECD. Retrievd from https://www.oecd.org/education/school/attractingdevelopingandretainingeffectiveteachers

Perry, T., & Delpit, L. D. (Eds.). (1998). *The real ebonics debate: Power, language, and the education of African-American children*. New York, NY: Beacon Press.

Pitts, D. W. (2007). Representative bureaucracy, ethnicity, and public schools: Examining the link between representation and performance. *Administration & Society*, *39*(4), 497–526.

Schleicher, A. (2018). *Valuing our teachers and raising their status: How communities can help*. Paris: OECD.

Schmitz, S., Nourse, S., & Ross, M. (2012). Increasing teacher diversity: Growing your own through partnerships. *Education*, *133*(1), 181–187.

Stokes, D., Evans, P., & Craig, C. (2017). *Developing STEM teachers through both informal and formal learning experiences*. Ediciones Universidad de Salamanca. Retrievd from http://repositorio.grial.eu/handle/grial/919

Texas Education Agency. (2017). *Out of field teaching report, 2016-2017*. Retrievd from https://tea.texas.gov/Reports_and_Data/Educator_Data/Educator_Reports_and_Data/

Texas Education Agency. (2018). Teacher retention 2014-2018. Retrievd from https://tea.texas.gov/sites/default/files/Teacher%20Retention%20by%20Route%202014-2018.pdf

The White House. (2016). Computer science for all. Retrievd from https://obamawhitehouse.archives.gov/blog/2016/01/30/computer-science-all

Toven-Lindsey, B., Levis-Fitzgerald, M., Barber, P. H., & Hasson, T. (2015). Increasing persistence in undergraduate science majors: A model for institutional support of underrepresented students. *CBE-Life Sciences Education*, *14*(2), 1–12.

U.S. Department of Education. (2016). *2016 title II report: National teacher preparation data*. Retrievd from https://title2.ed.gov/Public/SecReport.aspx.

Walsh, R. (2013, November). Let's talk houstonization: A new meaning for a once-dirty word. *Houstonia*. Retrievd from https://www.houstoniamag.com/articles/2013/10/31/lets-talk-houstonization-november-2013

York-Barr, J., & Duke, K. (2004). What do we know about teacher leadership? Findings from two decades of scholarship. *Review of Educational Research*, *74*(3), 255–316.

FINAL WORDS

Cheryl J. Craig, Paige K. Evans and Donna W. Stokes

When Paige Evans and Donna W. Stokes began to create the physics portion of *teach*HOUSTON and later when *teach*HOUSTON became funded by its first National Science Foundation (NSF) NOYCE grant, they had no inkling that a series of future grants would support the ongoing development of the program. Also, the research line developed with Cheryl Craig was not fathomed. Even when this book was proposed and approved as the first of a three-volume set, there was no grand, lock step plan in sight. Instead, the authors envisioned ongoing, unfolding productivity. As we look back, the three of us marvel at *teach*HOUSTON's progress, which has not been arrow straight to be certain, but mostly forward moving in the best sense of the term.

Surveying what has been accomplished thus far, we cannot help but remember Philip Jackson (1968) who considered his book, *Life in Classrooms*, a "mélange," saying it was a little bit about this and a little bit of that. But when everything was seamed together, it constituted Jackson's groundbreaking *Life in Schools.* Borrowing Philip Jackson's metaphor, our book is also a mélange. Parts of it illuminate this nut and that bolt and argue each piece's significance to the *teach*HOUSTON program and urban science teacher education in general. But then there are the serially interpreted chapters that provide overarching insights into the input of parents, the influence of teachers, the positive effects of certain professors, and the overall value of NSF scholarships to individual students. Also, the two ever-present mission threads allow our varying contributions to this book to cohere: science as inquiry and inquiry into inquiry – or more specifically the use of narrative inquiry to study science as inquiry.

From the beginning, the Physics by Inquiry course was pivotal. In many ways, it proved that physics could be both taught and learned through doing and thinking in action, which contrasted with some of the status quo science, technology, engineering, and mathematics (STEM) teaching pedagogies in the discipline-based colleges. The ongoing success of that course gave rise to other courses taught in the inquiry vein (i.e., biochemistry as inquiry). It also led to the Physics Department's acceptance of the physics education courses, which

Preparing Teachers to Teach the STEM Disciplines in America's Urban Schools
Advances in Research on Teaching, Volume 35, 227–230

ISSN: 1479-3687/doi:10.1108/S1479-368720210000035013

was no minor matter as we shared earlier. Overcoming this hurdle helped build a solid foundation for other grant applications and more grant activities that followed. It also established scientific inquiry as a vital thread and foundation around which *teach*HOUSTON was built. Summer activities, for example, all bear science-as-inquiry markings, as do all formal and informal learning tasks. Every submitted grant application has had inquiry at its core.

As for narrative inquiry, it fits hand in glove with science taught as inquiry. This should not be surprising, given that both concepts find their roots in Joseph J. Schwab's groundbreaking work at the University of Chicago. In the early 1960s, Schwab authored one of the leading papers in the field of science, *What do scientists do?* (Schwab, 1960/1978; also Schwab, 1962). In it, he synthesized the findings from 4,000 articles. That work remains "the best exposition that [the field] has…and…still serve[s] as a model of what science is and how it should be taught" (DeBoer, 2014, p. 2454) via the inquiry method. Where narrative inquiry is concerned, it was created by F. Michael Connelly, a student of Joseph J. Schwab, and D. Jean Clandinin, a student of F. Michael Connelly (with Cheryl Craig being a student of both of them) (Clandinin & Connelly, 2000; Connelly & Clandinin, 1990). Both creators of narrative inquiry were well-versed in stable and fluid forms of inquiry and presented narrative inquiry to the field as a fluid type of investigation. Also, Joseph J. Schwab gave them feedback in the development of narrative inquiry and declared their work as the next form of research studies needed in the field. It is important to add here that Paige Evans conducted her science as inquiry dissertation work using narrative inquiry as her research method with Cheryl Craig serving as her supervisor and dissertation chair. Intergenerational understandings are also at work in this science as inquiry and narrative inquiry union. This strand of research carries on with graduated doctoral students Michele Norton (engineering education/narrative inquiry) and Xiao Han (technology teacher education/narrative inquiry), for example.

Equally important to *teach*HOUSTON's development – if not more important – was the collaboration struck between the physics education professor (Paige) and the physics professor (Donna). Their mutuality was unquestionably foundational to the *teach*HOUSTON physics program that they shaped. The reciprocal relationship that emerged ensured that the program's development was foremost in mind. Egos were left at the back door so the front door could be open to following inquiries where they led in both scientific and educational research terms.

We cannot end this book without reflecting on our own career developments, which were propelled in some ways by *teach*HOUSTON's prominence on the state and national scene. First, Paige Evans began as a Master Teacher and, through hard work and commitment, is now a Clinical Full Professor, Associate Director of the *teach*HOUSTON program, and Executive Director of the *teach*HOUSTON STEM Interactive. Second, Donna W. Stokes has advanced from her Associate Professor in the Department of Physics to a Full Professor and the Associate Dean of Student Success and Undergraduate Affairs position at the University of Houston. Third, Cheryl Craig moved to Texas A&M

University from the University of Houston and became a Professor and the Houston Endowment Endowed Chair of Urban Education. She also is the Program Lead for Technology and Teacher Education there. Hence, it is safe to say that as we enhanced and promoted *teach*HOUSTON (each in her own way), *teach*HOUSTON assisted us by fueling what we could know, do and be, while concurrently advancing our careers.

Lastly, we want to turn to an important question that everyone seems to ask. Where does one begin to create a STEM teacher education program capable of combating the urban STEM teacher education crisis? We recommend that you seriously pursue answers to the questions you have personally framed. Along the way, we hope this volume could be used as an exemplar – a narrative of inquiry, so to speak – of a program of action, teaching, and research that you could potentially initiate. However, *teach*HOUSTON can never be a "rhetoric of conclusion" (Schwab, 1961/1978, p. 242) of what secondary urban STEM teacher education should be – as there is always "more to know and more to know about" (Schwab, 1956/1978, p. 153) as the program development of *teach*-HOUSTON reveals and this book writing journey has makes abundantly clear.

AUTHORS' NOTE

This research was funded by the following grants:

- National Science Foundation DUE Award 1240083, *Recruitment, Preparation and Retention of STEM Students as High School Teachers*
- National Science Foundation DUE Award 1557309, *University of Houston: Learning through Informal and Formal Experiences* (UH-LIFE)
- National Science Foundation DUE Award 1557273, *Collaborative Research: Understanding Robert Noyce Teacher Scholarship Outcomes in Texas*
- National Science Foundation DUE Award 1759454, *Leadership through Equity and Advocacy Development* (UH-LEAD)
- National Science Foundation DUE Award 1950036, *Preparing STEM Teachers through Advancing Cultural and Computational Engagement in STEM Scholars* (UH-ACCESS)
- Department of Education Supporting Effective Educator Development (SEED) Grant, U423A180139, *Expanding and Strengthening the STEM Teacher Workforce Through UTeach*

REFERENCES

Clandinin, D. J., & Connelly, F. M. (2000). *Narrative inquiry: Experience and story in qualitative research*. San Francisco, CA: Jossey-Bass.

Connelly, F. M., & Clandinin, D. J. (1990). Stories of experience and narrative inquiry. *Educational Researcher*, *19*(5), 2–14.

DeBoer, G. (2014). Joseph J. Schwab: His work and his legacy. In M. Mathew (Ed.), *International handbook of research in history, philosophy and science teaching* (pp. 2433–2468). Dordrecht: Springer.

Jackson, P. W. (1968). *Life in classrooms*. New York, NY: Holt, Reinhart & Winston.

Schwab, J. J. (1956/1978). Enquiry and the reading process. In I. Westbury & N. Wilkof (Eds.), *Science, curriculum and liberal education: Selected essays* (pp. 149–163). Chicago, IL: University of Chicago Press.

Schwab, J. J. (1960/1978). What do scientists do? In I. Westbury & N. Wilkof (Eds.), *Science, curriculum and liberal education: Selected essays* (pp. 184–228). Chicago, IL: University of Chicago Press.

Schwab, J. J. (1961/1978). Education and the structure of the disciplines. In I. Westbury, & N. Wilkof (Eds.), *Science, curriculum and liberal education: Selected essays* (pp. 229–272). Chicago, IL: University of Chicago Press.

Schwab, J. J. (1962). The teaching of science as enquiry. In J. Schwab & P. Brandwein (Eds.), *The teaching of science* (pp. 3–103). Cambridge, MA: Harvard University Press.

INDEX